Advance praise for BLUR

"Davis and Meyer catch the fundamental shift in how we all must look at our businesses, markets, and customers as we move forward in a more connected world. Their ideas are insightful, to the point, and essential."
—John Connolly, CEO, Mainspring

"A fascinating book that will cause you to rethink the next round of changes that will drive management into the twenty-first century!"
—Glen Urban, Dean, Sloan School of Management, MIT

"BLUR is fast, smart, and useful—a decoder ring that any business person can use to make sense of the turbulence in the world of work today."
—Alan Webber, Founding Editor, *Fast Company*

"BLUR is not another book about technology, but a comprehensive account of how the economy will change and how technology changes create new markets. Davis and Meyer provide original insights into the radical implications for the way we make decisions—about strategy, investments, products, and people.
—John Petrillo, Executive Vice President of Corporate Strategy and Business Development, AT&T

"An insightful and early probe into the new economy."
—Kevin Kelly, Executive Editor, *Wired*, and author of *Out of Control*

OTHER BOOKS BY STAN DAVIS

Future Perfect (1987, 1996)

2020 Vision (1991)

The Monster Under the Bed (1994)

BLUR

the
speed of
change
in the
connected
economy

Stan Davis

Christopher Meyer

CAPSTONE

Copyright © Ernst & Young 1998

First published in Great Britain 1998 by
Capstone Publishing Limited
Oxford Centre for Innovation
Mill Street
Oxford OX2 0JX
United Kingdom
Reprinted 1998, 1999

http://www.capstone.co.uk

Published by arrangement with
Addison-Wesley Publishing Company,
Reading, MA, USA

Jacket design by Suzanne Heiser
Text design by Dede Cummings
Set in 11 point Goudy by Vicki L. Hochstedler

British Library Cataloguing in Publication Data
A CIP catalogue record for this book is available from the British Library

ISBN 1-900961-71-7

Printed and bound in Great Britain by
T.J. International Ltd, Padstow, Cornwall

This book is printed on acid-free paper

TO ADAM, ALEXANDRA, AND SAM

TO MARY

Contents

IV. THE BLUR OF RESOURCES

V. LIVING THE BLUR

FOREWORD

This work is much more than a book. It is also a window into a conversation we hope you will join—an ongoing, organic exchange of opinions among people with differing views and experience. We are not offering the ultimate word on our topics, but a starting point: provocative ideas, observations, and predictions to get you to think creatively about your business and your future, and, we hope, to get you to share your thoughts with us. So most important, the book represents an invitation. We want you to join this conversation, take ownership of it, and help us take it to the next level of usefulness.

Before you read any further, take note of our Web site: www.blursight.com. Your experience of BLUR, shared via a Web community of interest, will enrich us all.

Interaction, in fact, is how BLUR began. In 1988, when Stan was
writing *2020 Vision,* the two of us started trading observations about
the transformation of the economy. We clipped articles, forwarded
e-mails, and as often as possible, sat down, reported on conversations
with others, and argued about what it all meant. Over time, our conversation about the information economy came to dwell on three
major themes: Connectivity, Speed, and Intangibles. Later, we drew
in more people. The conversation was particularly enriched when
Chris joined the Ernst & Young Center for Business Innovation^SM, as
its director. (Stan became a research fellow at the CBI, while maintaining his independence as an author and public speaker.) This
added not only a group of fine-minded individuals who contributed
their thoughts and a research infrastructure in which to pursue them,
but access to the experiences of E&Y clients. Chris Christensen, the
leader of the firm's Knowledge-Based Business practice, became the
third major participant in the conversation, and his field experience
helped ensure that our inferences and speculations had meaning for
managers and their companies. Terry Ozan, the vice chairman of
U.S. Management Consulting, was also a major contributor and
agreed to support the research.

Since then, everything has felt like a blur. We were surprised
how quickly, once many minds were contributing, the patterns of the
new economy revealed themselves. In this case, many cooks, it
became clear, *improved* the broth. And that's when we started to
become greedy. We realized we could get smarter still about the new

economy if we pulled even more people into our conversation. This book, therefore, is a means, not an end—a briefing on where we are so far. If you want to be generous, view it as a manifesto. It lays out a point of view, and the threads of a discussion, and suggests a new set of rules for managing. If you do nothing but read it, we're confident you'll find it of value. But if you're like us, you'll get much more out of the conversations it provokes.

We know that's where the value lies for us—in your ideas and engagement. We need you to be critical, to make us more rigorous, and to give us more and better examples of BLUR phenomena. Just as Netscape releases beta versions and asks users to contribute to refining the product, we're asking you to help us better understand what we've written. Like Netscape's Bugs for Bounty program, we'll compensate you for helping us to think more clearly about BLUR, partly financially, partly by access to the community we hope BLUR will create, and partly by helping you BLUR your own situation. See the Web site for details.

Finally, before you join this conversation, we want to make you aware of two more of its major participants: Julia Kirby and Nikolas Kron at the CBI have done triple duty, adding their insights to the conversation, turning up key facts, and keeping this volume on track. It has been their extraordinary effort that has enabled this book to be created at truly BLUR speed. We'll run the rest of the many other credits, movie-style, at the end.

Let's get to the BLUR.

An economy uses resources to fulfill desires.

An "economy" is the way people use resources to fulfill their desires. The specific ways they do this have changed several times through history, and are shifting yet again—this time driven by three forces—Connectivity, Speed, and the growth of Intangible value.

Because we are so newly caught up in the whirlwind of this transition, we are experiencing it as a BLUR. The BLUR of Connectivity, as players become so intimately connected that the boundaries between them are fuzzy; the BLUR of Speed, as business changes so fast it's hard to get your situation in focus; and the BLUR of Intangible value, as the future arrives at such a pace that physical capital becomes more liability than asset. Increasingly, value resides in information and relationships— things you can't see at all and often can't measure.

The arrangements we are all used to, like working for money, paying for goods and services, and maintaining clear boundaries between one organization and another are all blurring. Part I, **Seeing the BLUR,** is the beginning of understanding what will emerge in their place.

PART I

Seeing

the

BLUR

CHAPTER 1

PUTTING
BLUR IN FOCUS

Speed × Connectivity × Intangibles = BLUR

Speed:	Every aspect of business and the connected organization operates and changes in real time.
Connectivity:	Everything is becoming electronically connected to everything else: products, people, companies, countries, everything.
Intangibles:	Every offer has both tangible and intangible economic value. The intangible is growing faster.
BLUR:	The new world in which you will come to live and work.

Has the pace of change accelerated way beyond your comfort zone? Are the rules that guided your decisions in the past no longer reliable? If so, you are just like everyone else who's paying attention. You're not imagining things.

The elements of change that are driving these momentous shifts are based on the fundamental dimensions of the universe itself: time, space, and mass. Since the economy and your business are part of the universe, time, space, and mass are the fundamental dimensions of them as well. Until recently, this notion was too abstract to be very useful. Now, we are realizing the extraordinary power this insight has for the business world.[1]

Almost instantaneous communication and computation, for example, are shrinking time and focusing us on Speed. Connectivity is putting everybody and everything online in one way or another and has led to "the death of distance,"[2] a shrinking of space. Intangible value of all kinds, like service and information, is growing explosively, reducing the importance of tangible mass.

Connectivity, Speed, and Intangibles—the derivatives of time, space, and mass—are blurring the rules and redefining our businesses and our lives. They are destroying solutions, such as mass production, segmented pricing, and standardized jobs, that worked for the relatively slow, unconnected industrial world. The fact is, something enormous *is* happening all around you, enough to make you feel as if you're losing your balance and seeing double. So relax. You are experiencing things as they really are, a BLUR. Ignore these forces and BLUR will make you miserable and your business hopeless. Harness and leverage them, and you can enter the world of BLUR, move to its cadence and once again see the world clearly.

What will you see? A meltdown of all traditional boundaries. In the BLUR world, products and services are merging. Buyers sell and sellers buy. Neat value chains are messy economic webs. Homes are offices. No longer is there a clear line between structure and process, owning and using, knowing and learning, real and virtual. Less and less separates employee and employer. In the world of capital—itself as much a liability as an asset—value moves so fast you can't tell stock from flow. On every front, opposites are blurring.

With this book, we offer a lens through which you can see BLUR for a moment of static clarity, much as strobe-lit people seem to freeze in poses on a dance floor. Don't think you'll ever slow down BLUR, let alone bring it to a halt. Its constant acceleration is here to stay, and those who miss that point will miss everything. Your job as a manager, as an entrepreneur, as a consumer, and as an individual is to master the BLUR, to keep the acceleration going, to keep your world changing and off balance. Stop trying to slow it down. Stop trying to clarify it, codify it, explain it. Recognize it. Learn its new rules. You'll then be able to move at BLUR's speed—and discover that you can thrive in amazing new ways.

You're ready to join the dance on the other side of the strobes, where you'll move at the speed of BLUR. Remember, the frozen image is false. The reality is continual motion, a BLURstorm.

Connectivity, Speed, and Intangibles: The Trinity of the BLUR

We may tell ourselves that Connectivity, Speed, and Intangibles hold no surprises, but each has gone through its own metamorphosis in recent times.

On the evening of January 17, 1991, Chris was flying home from Los Angeles to Boston when the pilot announced that Operation Desert Storm had begun. He then broadcast the audio portion of CNN's report over the DC-10's entertainment system, so that the passengers could listen to a description by a reporter in Baghdad, relayed through Israel, transmitted worldwide, picked up by the airplane, and sent to their earphones. Given the importance of the event, what would you do? Within moments, Chris called his wife, Mary, in Boston to tell her the news. This all happened through an airborne-connected network that was far more powerful than anything that military pilots had in the Korean War, just 40 years earlier. How was this possible? By connecting everything to everything, in real time.

In rapid succession the deregulation of telecommunications, the miniaturization of satellites, and the development of mobile technologies have made connection available to anyone, anytime, any-place. Now, with the explosive takeoff of the Internet, we've entered the second half of the information economy, which uses the computer less for data crunching and more for connecting: people to people, machine to machine, product to service, network to network, organization to organization, and all the combinations thereof.

In 1999, low-earth–orbiting satellites (LEOs) will bring these capabilities everywhere on the planet. By the year 2000 telecommunication networks in the United States will carry more electronic data than voice. More astonishing, voice will be less than 2 percent of the traffic by 2003.[3] *Think of telephones running on the Net, rather than the Net running on the phone system!* The cost of global telecommunications will fall drastically, just as long-distance costs have done over the past 20 years. This massive increase in connection will change the way all business is conducted, turn current business and economic models on their heads, and take us ever deeper and faster into BLUR.

Mobile phones, pagers, voice mail systems, bar code scanners, satellite phones, e-mail, global positioning satellites, and all the rest of the electronic gizmos that will connect us are just the visible part of the Connectivity story. Once these things connect with each other, their actions will trigger domino effects and change the way the economy behaves.

The stock market crash of October 1987 was caused by computerized trades, none of which were linked explicitly. The damage was done by the interaction of independent investor instructions—a kind of connected network of trading programs. Similarly, the 1965 Northeast power blackout was the worst in history because of the connections shared by the utilities in the power grid. These connections translated an overload at one point in the system into a cascading failure throughout the grid. That's why the trading limits installed after the 1987 financial debacle are called "circuit breakers."

These examples illustrate the importance of connection: A single automatic circuit breaker, a cruncher, has limited impact. Connect a thousand of them, and there's no telling how the system will behave. The first use of the financial circuit breaker, in October 1997, was in response to frenzied selling in markets in Asia earlier in the day, further evidence of the connectedness of the global system.

The instant communication of the Gulf War is an amazing story about Connectivity, but a trivial one about Speed. News has always traveled fast. When we speak of Speed, we're talking about how the sheer velocity of business has increased over the last decade. Take, for example, the dramatic drop in the time between sending and receiving. Order a laptop from PC Connection at 2:00 A.M., receive it custom-configured at home ten hours later. Similarly, the time lapse is evaporating between producing and selling, and purchasing and delivering. In Tokyo, you can order your customized Toyota on Monday, and be driving it on Friday.

Speed is the foreshortening of product life cycles from years to months or even weeks. And, Speed is the worldwide electronic network over which financial institutions transfer money at the rate of $41 billion a minute. For the individual, Speed is scores of messages a day, creating near continuous communication. Miss a day and your world moves on without you. Accelerated product life cycles and time-based competition have become part of the business lingo. These experiences change people's perceptions. We now expect real-time responsiveness, 24 hours of every day of the year. This premium placed on anytime, real-time responsiveness is just one example of the growing importance of intangible value.

The intangible portion of the economy has grown quietly, altering how we see the world without calling too much attention to itself. It takes four forms. *Services*, which have dominated the economy for decades, are the most familiar. They include everything from hamburger flippers to brain surgeons. The second is *information*, like the specialized knowledge in databases, the content of George Lucas's Star Wars empire, and the shadows cast by activities and transactions onto credit card bills and stock tickers. The third is the *service component of products*. Computer simulation services, on everything from automobiles and architecture to windows and wallpaper, commonly allow customers to try out different configurations before ordering the actual product. Fourth are *emotions*, the trust and loyalty that people feel for a brand, the prestige conveyed by a label, the attraction exerted by a celebrity on the stage, screen, or playing field.

DreamWorks SKG is a powerful expression of the value of Intangibles. Formed with a mere $250 million, it was mobbed by eager investors who drove the market capitalization to $2 billion when the company, which had no studio and nothing in production, went public. The attraction was the intangible value of its founders: Steven Spielberg, Jeffrey Katzenberg, and David Geffen, each of whom is a star in the entertainment business even though their faces are never on the screen. Intangibles are at work in the form of innovation, brands, trust, and relationships.

Intangible value is growing much faster than the tangible. But in the connected economy, the increasing demand for Intangibles also brings with it some new economic wrinkles. Compare succotash and software. With succotash, two pounds cost twice as much as one pound. But with software, the second copy is pretty much free. Further, if we cook a pound of succotash, we eat it *or* you do. Software, however, can be duplicated again *and* again with no additional cost.

Connectivity, Speed, and Intangibles are keeping all of us up nights, thinking about our future. To see where they will take us, focus on the constant of what an economy does.

An Economy Uses Resources to Fulfill Desires

The only constant is what economies do: again, they use *resources to fulfill desires*. Their means of doing so, however, have changed several times. Hunting-and-gathering economies lasted about one hundred thousand years before they gave way to agrarian economies, which endured 10,000 years. Their labor- and land-intensive approach was succeeded by the machines and factories of the Industrial era (1760s–1950s), which spawned the growth of cities, mass production, pollution, labor unions, and the development of the banking system.

After almost 200 years, the industrial era gave way to the computers of the information economy, which is already half over. The first four decades used the computer as a crunching tool, an industrial-style approach that included data processing and warehousing, bigger and faster machines, supercomputers, and other "factories" to perform routine brain work.

In this second half of the information era, which we call the BLUR economy, resources will fulfill desires by yet another set of arrangements. One will be that we leave behind the idea of stable solutions. Already, a successful business is neither at rest nor in focus at any given moment. The beginning, middle, and end of a product line are dissolving into each other as the orderly and familiar step-by-step progression of research, design, production, distribution, payment, and consumption disappear. Advance copies of the manufacturers' new models are being reviewed in magazines while the current model is still being sold at a floating "street" price, and the previous version is available at a discount. An annual model change blurs into a continuous one. Continuous upgrades that are downloaded electronically are replacing model years that require plants to close down for retooling. *Built to last now means built to change.* Like grandfather's rocking chair, all the pieces may have been replaced over the years, but the intangible—the concept of grandfather's chair—persists.

Like those before it, the BLUR economy has three basic parts:

1. The BLUR of desires—the demand side of an economy—where products and services meld into one to become an offer, and where the roles of buyers and sellers merge into an exchange.

2. The BLUR of fulfillment, where strategies and organizations dissolve into economic webs and permeable relationships.

3. The BLUR of resources, where people are no longer divided into their working and consuming selves, and where capital is more often a liability than an asset. These resources are shaking off their traditional meanings as vigorously as a dog shakes off water after climbing out of a lake.

How to Read and Use this Book

The very nature of BLUR means that all three parts are connected and must be considered simultaneously. You can read the parts of this book in either direction: resources, fulfillment, desires; or desires, fulfillment, resources. We've organized it to start out in the market with The BLUR of Desires, but you can also read BLUR beginning with People and Capital. True to BLUR, Stan likes to read the book in one direction and Chris in the other!

Unlike most business books, BLUR is not going to identify problems and recommend solutions. That's fixing yesterday's concerns, which would only be investing in your weaknesses. Instead, we hope to provide a motion-altering camera lens through which you can see how the convergence of Connectivity, Speed, and Intangibles is creating BLUR. We want to redefine how you think about your place, and the place of your business, in the economy.

By the end of this book, you should be equipped to see beyond today's snapshot of all that's happening. Better yet, we hope you'll be able to embrace it and join it. In the last chapter, we give you 50 ways to BLUR your business and 10 ways to BLUR yourself, fast examples of how you can apply what you've learned. Remember, even after you set down this volume or quit your PC or Mac, the dance continues at full pelt and the images go on changing. It's already well past what we've assembled here. We must all keep moving, creating a community of BLUR. Come join the dance.

Desires

Human needs have changed little throughout our economic history, and are unlikely to change now. But the ways in which needs were traditionally fulfilled in the Industrial era trained most of us to think in terms of products and services, and buyers and sellers. This is no way to think about the future.

Part II contains two chapters. In The Offer, we examine economic desires to show how products and services are blurring into offers that fulfill broad desires, rather than becoming components for assembly by the customer. In The Exchange, the distinct roles of buyer and seller on opposite sides of a transaction are broadened to define a two-way trade in which both buyer and seller exchange value of three kinds: economic, informational, and emotional. Together, these chapters describe the transformation of the "demand side" of the economy.

PART II

The BLUR of Desires

CHAPTER 2

THE OFFER

No Product without Service
No Service without Product

The difference between products and services
blurs to the point that the distinction is a trap.
Winners provide an offer that is both product and
service simultaneously.

Imagine driving down the road in your Mercedes when you hear a faint noise. Is it pinging? Is it knocking? (Is it the belt of your raincoat caught in the door?) What are the chances it will do it again if you take it into the shop? Don't worry. Mercedes is testing a new system that will connect the car's software via the Internet to a customer assistance center. This will enable them to diagnose problems while a car is still on the road—and sometimes even download the repair.

Hilton Hotels allows guests to bypass the registration desks at some of its establishments and to check in at "smart card" kiosks instead. In a pilot program being conducted in conjunction with American Express and IBM, a traveler's AmEx card stores her preferences—nonsmoking, king-size bed, whatever—along with her Hilton HHonors number, and allows her to update her "profile" at any time. Then the kiosk spits out a key, along with directions to the room.

When a car owner buys a LoJack antitheft device, there's really no assurance that the vehicle won't get stolen. But it's almost a certainty that it will be recovered quickly if it is. The hidden device radio-signals its exact location to police cars in the vicinity. For $595, the owner gets peace of mind—and the last laugh. The next generation may well send a signal to disable the stolen car.

Cars, hotel rooms, security systems: Are these things products or services? In a world of Connectivity, Speed, and Intangibles, not only is it hard to tell, it's not particularly productive to try. In all these cases, as in numerous others, the once clear line that separated the two is disappearing, an evolutionary process that is at the heart of BLUR.

This makes sense given the forces behind BLUR. Speed—the acceleration of business in every respect—is driving products and services to resemble each other in various ways. For one, product life cycles are a fraction of what they used to be. In turn, rapid obsolescence demands continual upgrading or replacement. There's no such thing anymore as selling a product to a customer and then forgetting about him. The people who are your customers today will be customers again in six months—if not yours then someone else's. When you're dealing with the same customers with that frequency, doesn't it begin to qualify as a service business?

Conversely, the need for Speed is causing a lot of people in the service business to feel as though they're selling products. Customers want their services *fast*, whether they're buying a burger or financing a home. Under pressures of time, there's often no way services can be totally tailored for each customer. At their core, even when services are mass customized,[1] they still have to be standardized, modularized, packaged, and embedded in software. In other words, they have to look a lot more like products.

Accounting for even more product/service confusion is the growing proportion of business value that is made up of Intangibles—

which of course, are what service is all about. The concept of service, after all, is that you are not buying or selling something to have and to hold. Nevertheless, Intangibles are beginning to incorporate the essence of what traditionally have been labeled products. In a product such as Tylenol, for example, the biggest value component is hardly the few cents' worth of active ingredient. Rather, it's the scientific expertise that went into the drug's seven-year development. In an Ebel watch, it's not the cost of materials that drives the $1,000 price tag. It's the talent of Ebel's designers and the prestige of the brand.

Chiefly, though, products and services are blurring because of the prime BLUR force, Connectivity, the ability to build products that link electronically to information bases. This creates many opportunities for service. Think back to the LoJack example. LoJack is a service more than it's a product. In fact, not only can't LoJack owners see the unit, they aren't even told where it's installed on their own cars. But the pricing sure sounds like a product. There's no monthly maintenance fee and the upfront price is the buyer's only cost. In short, Connectivity means that customers maintain a close link to the creators of the goods they use. The product is simply a service waiting to happen; the service is the product in action.

Clearly, it no longer makes sense to think of the world in terms of products and services. Instead, we should be thinking more broadly of product-service hybrids—"offers" in the blurred lexicon. Offers are "productized" services and "servicized" products. They're both fish and fowl, if you like. And more frequently, it will be hard to sell anything that doesn't represent that combination. Vendors who sell unsupported, unconnected products will be viewed as no better than snake oil salesmen, quick to decamp during the night and foist their wares on a group of unsuspecting buyers in some other town. Vendors who sell unpackaged and unleveraged services will be viewed as concerned only with running up their billable hours, and unconcerned with fulfilling the customer's desire. They will be undercut by the economies of more productized competitors. Avoiding these fates will require a whole new mind-set before planning what to offer. The successful design of offers will require thinking simultaneously about what it is, and what it does, and even what it enables to happen.

Beyond Bundling

As we'll find in all things BLUR, there are clear antecedents to the innovations we are seeing, but they pale in comparison to the changes to come. It has always been the case, after all, that services have been "bundled" with products, and vice versa. Buy a television or VCR and you will no doubt be given a service warranty and encouraged to buy an extension. The price of a car stereo often includes installation. Until recently, mail-order pioneer L.L. Bean offered "free shipping," which actually meant the cost of transportation was already bundled into the price you paid for the merchandise. Likewise, products are bundled into services. Hire a diaper service and the diapers are part of the deal. Go to a time-management course by the Franklin Planner people and they throw in the planner. Join Weight Watchers and learn why you should opt for the balanced, low-cal meals Weight Watchers offers at the supermarket.

Bundling is a step in the blurred direction. It demonstrates an awareness that people buy products and services because they have needs that these things, *together*, help to fill. Bundling more product or service into the original offer is a way to meet more of that need. That same good instinct has led to the widespread use of the '90s buzz phrase, "value proposition." Management thinkers urge product developers and marketers to tout benefits, not features. Presenting the customer with a value proposition means addressing what it is she is trying to achieve, and how the product or service will get her closer to that goal—not showing it off in its own right.

Ultimately, however, bundling isn't enough. It's not sufficient to toss together a set of complementary goods and services and stick a price tag on the lot. The real value comes when these things are blended so that they can't be separated—and can't exist independently of each other.

Offers We Can't Refuse

A few offers already in the marketplace are so innovative in their design and have proven so successful that you're probably already familiar with them even if you're not aware of it. Otis Elevators installs sensors in various working parts of its equipment to monitor usage and performance. An onboard computer equipped with a modem using cell-phone technology collects that information. A mechanical failure, or the threat of one, signals the computer to call Otis headquarters and a repairperson is dispatched immediately. Though such emergencies are rare, the constant monitoring makes preventive maintenance much more focused and effective. In addition, Otis technicians periodically check in remotely, to examine the elevators' various components and their state of wear. When it is time for routine maintenance, the mechanic comes armed with a very good idea of where degradation might be starting to occur. This way, "wrench time" can be focused where it's needed rather than giving every part equal attention. Service costs go down and service quality goes up. What an offer!

It wasn't always so. Otis used to have two separate businesses, one to build and sell elevators, the other to sell and deliver maintenance services. Integrating the two has been only partly a technology task. At least as difficult has been the training of employees to operate as required by the integration. Most of the physical blurring still lies ahead. Otis has more than 100,000 elevators in service in North America. It began installing the new system in 1997, and expects to be able to convert only 10,000 to 15,000 per year for the foreseeable future.

Another famous marriage of product and service took place at Dell Computers. Dell competes in a cutthroat business, offering rock-bottom prices on personal computers and related peripherals. Founder Michael Dell's strategy has always been to cut fat out of the supply chain as a way of keeping prices low. Dell's innovation, around 1987, was to sell directly to customers and avoid all the inventory and handling costs of stocking store shelves. What's powerful about this business model, though, is that it has enabled Dell to be more responsive to customers' individual requirements. Buying through more traditional channels, you would have to go to a store to choose a computer and specify the components and features you need—and then wait an average 14 days for delivery. Dell dropped this to a then-unheard of 4 days for a customized PC. Customizing each buyer's specs does not seem to present any particular difficulty. Practicing mass customization, the company waits to get the order before assembling the product, and ships in 2 days.[2] This transformed the business model for the industry. IBM, Compaq, and others connected to their distribution partners such as MicroAge or MacConnection in order to move final assembly downstream. This allowed them to compete with Dell—an example of how a demand for Speed led to greater Connectivity.

Dell's business has translated beautifully to the electronic commerce environment. Selling over the Internet has further sparked revenues by as much as $6 million a day, 80 percent of which are to new customers. "The Internet for us is a dream come true," says Dell. "It's like zero-variable-cost transactions. The only thing better would be mental telepathy."

Mass customization is the theme of another well-known product/service mix: Levi's Personal Pair™ program. You can go into a Levi's store and be fitted for a pair of tailored jeans—perfectly cut and sewn to your proportions, and possibly unique among the zillions of pairs Levi Strauss has turned out over the years. Your measurements are transmitted to Levi's cutting room, and your Personal Pair is mailed to you soon after. What's more, the online system retains your information in case you want to order a second pair, which you can do from any Levi store. The world's ultimate mass-produced product has been seamlessly integrated with service. (Your responsibility, of course, is to maintain your shape.) The same service is now available for footwear: For $275 and up, Digitoe will make a personal pair to your measurements, deliver them in three to four weeks, and keep your statistics on file for repeat orders.

Offers are emerging just as rapidly in what used to be the exclusive domain of service. Take tax preparation. Until recently, you could either prepare your own taxes or you could hire someone to do them for you. If your tax situation was at all complicated, and your own time valuable, you probably used a professional, someone who more than paid his way because of his knowledge of all the byways of the arcane U.S. tax code. But tax preparation isn't exactly a safe playground for creative spirits. It's a very disciplined task, informed by rigid methods and rules. The manufacturers of TurboTax, MacInTax, and others realized the process was a perfect service to embed in software and sell as a consumer product. Their software asks you all the same questions a service provider would, and applies the same level of knowledge instantly—without consuming the time of highly trained professionals, who are occupied instead building even more knowledge into their software.

On a much grander scale, Systemanalyse und Programment-wicklung AG, or SAP, essentially replaces traditional customized systems-integration services. SAP produces a suite of operations-management software that enables companies to integrate business activities upstream and downstream for greater efficiency and speed. For instance, it links raw materials purchasing to production planning to sales forecasting in a wide range of situations. Before SAP, companies had to custom-build systems that could do such work, or even more common, jury rig different links and connections among a legacy of disparate systems. SAP has changed what once was the work of a huge and growing systems integration service industry into a software-intensive product.

As these examples show, when services turn into offers that contain product, one common result is a shift toward self-service. You can be sure that all those folks toiling away at your neighborhood tax preparer's office always had the equivalent of a MacInTax to help them work with assembly-line efficiency. For Intuit, it was only a short step from there to put such a tool in the hands of the individual taxpayer, and remove the intermediary. Similarly, once banks had equipped their tellers with the information technology to do their jobs better, it became clear that the tellers' limited availability reduced value more than their personal service increased it. The bank could install an ATM into the side of the building and let the customers make their own withdrawals and deposits.

Andersen Windows provides another good example of the product/service BLUR. The company noticed that more builders were using nonstandard windows as a way to distinguish homes and offices. This meant they were hiring design and carpentry services, rather than going for the mass-produced goods that Andersen was known for. The company moved fast to head off widespread desertion by productizing those design and carpentry services. Now, a builder can sit at a PC and use a special Andersen program to design the perfect, unique window. When he is satisfied with how it looks, he simply transmits the order to Andersen to build it. Back in Baysport, Minnesota, the company has reconfigured its assembly processes to do this custom work efficiently.

One of the latest breakthroughs in self-service is E*Trade, the online brokerage service. Again, the great innovation is to dispense with the intermediary by putting the kind of technology used by brokers in the hands of individual investors. Citibank offers a similar capability at the corporate level. It offers clients a set of integrated software products that allows them to self-manage their investment portfolios. The bank helps set risk and pricing parameters that let clients proceed at their own speed, and builds in "advisor" toolkits to aid in decision making. Essentially, the individual investment advisor is following in the evolutionary steps of the bank teller.

All of the companies described here have succeeded in making offers to the market that capitalize on the blurred economy. Their products include services and their services are delivered like products. Each example contains guidance on how you can sell virtually anything today.

The 10 Attributes of an Offer

It's hardly surprising that many of the best offers we can point to are creations of the software industry. Software was born in the age of BLUR, and is one of its key enablers. It reduces the cost of service and increases its reliability. But as more of the items we regard as everyday goods are enhanced with software, it won't be just the software industry that looks blurry. Microchips and code are becoming ubiqui-tous. They're in our day planners, our hotel keys, our toasters, our cars, our pets. Did we say pets? Yes, in a variation on the LoJack theme, a Canadian company called PETtrac implants microchips in dogs and cats (4 million so far, and counting) as a way to help recov-er lost animals. The network, which involves veterinary offices and animal shelters, has already returned more than 17,000 pets to their owners. Will this spread to children, as a security measure? Possibly. Many more microchips are now going into noncomputer entities than into computers.

In 1975 the number of micro-controllers in computers equaled the number in noncomputers (everything else). Today the figures are far from equal. There are almost 9 times as many micro-controllers in noncomputers as there are CPUs in computers. These include approx-imately 9 billion microchips in modems, phones, pagers, automobiles, and consumer electronics products.[3]

As our basic consumer goods become increasingly software-intensive, we can expect that they, the world they make, and the indus-tries that produce them will start to behave in new ways. The chip will take over as "host object," and the toaster, for example, will take on the attributes of software. The point is that everything we build will be increasingly software-intensive with the result that all products will start behaving like those in the software industry,[4] which means they'll be fast and intangible. And as BLUR moves from the crunching period of the information age into the connecting era, they'll be networked, as well. The table on the next page shows the attributes of blurred offers.

ATTRIBUTES OF BLURRED OFFERS

Speed	Connection	Intangible
Anytime Customer Access and Response Real-time Operation	Online Interactive Anyplace Customer Access and Response	They Learn They Anticipate They Filter They Customize They Upgrade

Let's take a further look at all these attributes, keeping in mind that the most essential feature of all offers is that they are connected.

ANYTIME

Accessibility by users at any time of day is becoming a must-have for offers of all kinds. Who complains about bankers' hours any more? The bankers may still keep them, but their offers don't. At any time of day or night, account holders can check a balance, see if a given check has cleared, transfer funds, as well as, of course, deposit or withdraw money. The spectacular growth of the mail-order industry—and on its heels, electronic commerce—can be attributed largely to the public's hunger for anytime shopping. As more homes have both heads of family in the workplace, the demand for off-hours access to goods and services has risen—and with it, the expectation of "7 × 24" 800-number service. *Is your business available to customers any time?*

REAL TIME

This need for Speed of response in today's business environment puts a premium on systems that can operate in "real time." Rather than acting on historical experience or expectations about the future, these systems capture the reality of what's happening now, and function "on the fly." This capability has been an incredible boon to marketers, who otherwise must rely on traditional, asynchronous market research. Consider what Coca-Cola has done with its 800,000 vending machines in Japan. Normally, the machines are refilled and serviced by route drivers, who upon arrival see that a machine is out of Diet Coke, say, or Sprite. What is not known is how long ago each ran out, or whether there was a run on Sprite simply because frustrated Diet Coke drinkers opted for it as a second choice. New vending machines in Japan give Coke staff that kind of information in real time. Each is equipped with a microprocessor that communicates customer purchases back to a central distribution facility. As soon as the machine runs low on any particular soda, it calls for more. Coke has demand data for each location, by time of day—the ultimate in micro-market segmentation.

In the case of a new timepiece created by Seiko, real time translates literally to real time. The MessageWatch is equipped to receive satellite signals that automatically update it to atomic time (as established by the National Bureau of Standards) at least 36 times a day. The receiver on the watch can also collect assorted real-time information—stock market prices, for example—and display it on the watch face. *Can your business operate in real time?*

ONLINE

The key to capitalizing on the pervasive Connectivity of the BLUR
economy is to put your offer online. This keeps it connected all the
time to real-time information. A perfect example is PCS Health
Systems, the pharmaceutical benefits-management firm owned by
Eli-Lilly. PCS uses an online computer system to connect the 54,000
pharmacies and 50,000 physicians in its plan. Whenever a patient
presents a prescription, the software searches that patient's record to
check what other medications she has been prescribed in the past, by
any doctor in the network and regardless of where it was filled. The
software looks for drugs that are known to be dangerous in interaction with the current prescription. If it finds any, a red flag is raised
to the pharmacist, who then alerts the patient and the doctor. Is the
patient still taking the former prescription? If so, the new one may
cause trouble. In 1995 alone, the system flagged almost 45 million
such interactions, 5 million of them potentially fatal. *Is your offer
online?*

INTERACTIVE

Another great benefit of online systems that eliminate the middle person—travel agent, bank teller, stockbroker (and someday maybe even
the physician)—is that they can easily be made interactive. Think of
your favorite Web site. The ability it gives you to view information
selectively, pick different pathways, and ask—and get answers to—all
kinds of queries is a big part of its appeal. A typical, for-profit Web
site is CDNow, which allows music lovers to order virtually any

recording commercially available. A primitive version of the site might simply have presented an alphabetical listing for scrolling. Today, the site is highly interactive. Customers can look up recordings by composer, name, style, artist, label, or other criteria. Barnes & Noble's site does the same for book lovers.

Online offers aren't the only beneficiary of interactivity. Some offers communicate much more locally—indeed, there are those that talk only to themselves. Sunbeam's "blanket with a brain," for instance, features a thermostat mechanism that tells it when to send more heat to the feet and less to the shoulders. Miele, a German appliance maker, has built interactivity into its self-cleaning ovens. An electronic control unit senses the amount of smoke in the oven, the purity of the air, and the degree of soil on the surfaces, and automatically controls the duration of the cleaning program accordingly. Less warm-and-fuzzy are today's "smart guns" that will fire only for their legitimate owners. In one such system, a radio signal creates interactivity between a wristband and the trigger mechanism. Without that link, the weapon won't fire. *Is your offer interactive?*

ANYPLACE

Hand in hand with anytime access goes anyplace access; this is the other half of the mail-order boom. Overworked shoppers love that they can order everything from pantyhose to spiral-sliced ham from a comfortable chair in their living room—or from an airborne phone 31,000 feet above ground. Of course, the telephone, and now the Net, make ubiquitous access possible for Intangibles, and for many of the service aspects of almost all offers. The trick is being able to service anyplace access effectively—anytime. *Is your offer available to customers wherever they are?*

LEARNING

Offers really start to get interesting when they make it possible to learn; that is, when they can not only capture information about their use, but make adjustments or initiate action in line with that new information. Here's an example: On the 1997 Mercedes Sport Utility vehicle, the five-speed transmission is "driver-adaptive," meaning it revises the shifting speeds and conditions in accordance with the driver's habits. The capability goes beyond being merely programmable, like eight-way adjustable seats, because the offer itself—not its owner—makes the adjustment. It also goes beyond having memory (the "previous channel" key on your remote doesn't count). It has to do with active pattern recognition.

Have you ever received a call from your credit card issuer checking on a recent transaction? When it's done right, such a check happens because the purchase lay outside the behavior pattern you've established over time. The issuer's software has learned your habits, and knows an aberration when it sees one. Learning will become an even more prevalent feature of purchasing plastic as more *smart cards* come online. One species of smart cards is essentially a debit card capable of housing a great deal of information about its holder. Each contains a micro-controller with 8 or 16 kilobytes of memory, plus the usual 200-byte magnetic strip. Each time a smart card is inserted into a retailer's terminal, the machine collects data about the buyer and his purchases. Inevitably, if customers don't push back, the banks that issue these cards will start to sell the information the chips collect. This, in turn, will put many other offers on the learning curve. *Does your offer get smarter with use?*

ANTICIPATING

Once offers have the ability to learn, it's just a short step to give them something even more blurred: the ability to anticipate. Offers that anticipate can extrapolate from the patterns they've been able to observe, prepare for, and sometimes even suggest, what should probably come next. Thus, Web-site software such as Firefly or Net Perceptions is able to look at a user's past purchases, compare them to those of its broader population, and make suggestions of what the user might like to see or purchase next. Firefly is in use on sites such as BarnesandNoble.com (for book lovers) and myLAUNCH.com (for music lovers). MyYahoo!, meanwhile, offers personalized recommendations of Web sites based on users' preferences and makes it possible for those users to meet other people with similar tastes and interests. In short, it has productized editorial service! *Does your offer anticipate your customers' needs?*

FILTERING

A special form of customization is the filtering of the wide range of information and choices that increasingly confront users. Filtering is critical in many situations, though probably never more so than for jet fighter pilots. Given the speed and complexity of these aircraft, no pilot can deal with the torrent of incoming information in real time. Instead, smart systems help to filter out the "noise"—or extraneous, will-keep-till-later stuff—in the data, and release only that which is relevant to what the pilot is trying to do—survive.

Back on the ground, filtering is the chief selling point of Point-Cast, a personalized online news source. Its users consciously select filters to limit the news they receive to subjects of interest. Another filtering offer, CDNow, asks registered users to state whether they are, for instance, fans of fusion jazz or Baroque music. It then sends them e-mail messages to inform them of new releases in their areas of interest.

Filtering, by the way, is one of the most controversial topics in the Internet arena right now—largely centered on the Platform for Internet Content Selection, or PICS. PICS is a proposed standard that would make it easy for any group to offer its own filter on the Web. Thus, the Anti-Defamation League might be able to ensure its members surfing the Web that they won't run across anti-Semitic content. But by the same token, the Chinese government could use filtering as a way to censor news, blocking its citizens from learning of independence struggles in Tibet, for example. Filtering is clearly a double-edged sword. But in the world of BLUR, it will undoubtedly be a more common feature of commercial offers. *Does your offer deliver only the information your customers desire?*

CUSTOMIZING

Customization is a major theme running through the offers cited so far, whether they involve computers, jeans, or books. PointCast is the software example of the moment. It is designed to be customized to fit the precise news interests and needs of the individual buyer. Out of the box, PointCast asks you to give it its marching orders. Do you want news from certain cities or on certain topics? Which ones? Do you want to know the weather forecast? How frequently? And for which locations? How about tracking certain stock prices?

It's almost certain that no two copies of PointCast are delivering the same news. It's what writer Nicholas Negroponte envisioned when he predicted the "Daily Me,"[5] and is a far cry from your local newspaper or from anything that could be supported without a high degree of Connectivity. Customization is a BLUR phenomenon, and it is taking over. Soon it will be unheard of to market anything as "one size fits all." *Are your offers customizable?*

UPGRADING

One distinctive feature of software products is the constant stream of upgrades they spawn. Once you've bought a program, it isn't necessary to make a whole new purchase when its functionality is improved. The upgrade is offered at incremental cost, representing perhaps 20 percent of the original (or in cases such as Netscape, at no cost at all). The software underlying offers such as telephone and cable service make them similarly upgradeable. If you decide you want HBO or call-waiting, you don't need to turn in your cable box or phone. You just order the upgrade and someone somewhere flips a switch. And, as software permeates all kinds of goods and services, we can expect this to be the case even for very tangible offers. Imagine you hear about a new microwave oven that can detect when a liquid being heated has reached a boiling point and will shut off before it boils over. Nice feature, but is it worth throwing out the oven you bought last year and plunking down another three hundred bucks? Probably not. But what if that feature were offered as an upgrade, costing $9.99 and requiring no equipment replacement? When offers can be fulfilled simply by downloading upgrades, "model changes" will be only for styling and aesthetics or to take advantage of whole new platforms. Everything else will change constantly and incrementally. *Are your offers upgradeable?*

In summary, do all your offers in the marketplace have all 10 capabilities? If not, what are you waiting for? Get blurring!

Every Business Becomes a Software Business

People who design networked software had a tremendous advantage during BLUR's early days: They had no choice but to think in terms of offers; there was no constraining history of a product- or service-only orientation. For designers of other kinds of offers—those that have product or service antecedents—the challenge may be much greater. The purpose in outlining the 10 preceding attributes is to help jump-start the reconceptualization required by these more established businesses.

If you think that this revolution is only for high-tech products and services, consider the automotive industry. Evidence abounds that even the oldest of dogs can learn the new tricks. Automakers, those icons of the industrial age, have been hard at work creating knowledge-based products that are souped up with service. As Todd Lappin wrote in *Wired*:

> Peel away the slick, carbon-fiber skin of an Indy car these days, and you'll find yourself staring at a sophisticated nervous system of serpentine wires, precision sensors, LCDs, electronic black boxes, and spread-spectrum wireless communications equipment. Today's Indy cars take to the track equipped with two-way voice-communications systems and all the hardware needed to transmit reams of real-time performance telemetry back to pit row, where a new breed of race-crew technicians known as DAGs—short for Data Acquisition Geeks—use laptop computers to monitor the pulse of an 800-horsepower data terminal by remote control as it zips around the racetrack at more than 220 mph.[6]

And it isn't just race cars that implement this sophisticated technology that completes the BLUR between product and service. Most new vehicles carry a host of computer chips—in fact, a growing percent of their costs and perhaps as much as 90 percent of their value are estimated to be in their computers and the software they use. Some of these chips are devoted to establishing Connectivity and interactivity between different parts of the vehicle. Airbags, of course, deploy at the first signal of impact. Antilock brake systems kick in when sensors detect a spin. Cadillac has developed sensors that can take its cars out of a skid without help from the driver. The software compares the direction the driver is steering with where the vehicle is actually going. If it deduces the vehicle is out of control, it activates a traction control system to put it right. Toyota is taking safety a step further. On the Advanced Safety Vehicle the company is currently testing, onboard cameras and radar detect when the car is headed for a solid object. If the driver doesn't hit the brakes, the car sounds two warnings, then goes ahead and applies the brakes for him.

Many other innovations in the auto industry enable the vehicle to interact with the driver. Buyers of recent Mercedes models, for example, are told they don't have to adhere to a regular, miles-based schedule for oil changes and transmission fluid. Instead, the need for changes is determined electronically. Tires, too, have gained the power to communicate. Goodyear's Extended Mobility Tires, already available on Chevy Corvettes and Plymouth Prowlers, use a low-pressure sensor system to alert drivers when a tire is going flat. And how's this for customer intimacy: Toyota's new safety vehicle gives you a sharp poke in the back if it senses you're nodding off. (Such impoliteness is employed only as a last resort, of course. First, the warning system flashes a coffee cup icon on the dashboard, sounds noisy chimes, and shouts "wake up!" Only then does rude, physical contact occur.)

The next phase in turning autos into offers involves enabling their onboard software to interact with totally external entities. Vehicles featuring navigation systems can already link to satellite-based global positioning systems and employ comprehensive map databases. They display where you are at the moment, track your progress down the road, and issue directions to your destination. Coming next: Internet addresses for cars. Mercedes' Internet on Wheels prototype allows dial-up access between, for example, a car and a Mercedes service center. Data from the vehicle's onboard diagnostic software can be downloaded for analysis. Upgrades from the service center—a timing adjustment to improve mileage with a specific gasoline, for example—are just as easy to transfer. A similar capability is simultaneously being developed by Porsche in cooperation with Siemens. GM's OnStar Service is alerted whenever your airbag deploys, and calls your cell phone to see if you are safe. If either you or your phone is too broken to answer, OnStar calls for help.

The point of all these automotive examples is simple: If they can do it, you can do it. Automakers have recognized that people don't buy cars and trucks; they buy safe, convenient, and enjoyable transportation. They don't want just a product and they don't want just a service. So think about what you sell. Is it online yet? Which parts of it might be? How could it be made more interactive? How would the customer like it customized? If you planted a microchip in it, what kind of information could it collect? Go back to the list of attributes outlined earlier. How could your offer do all of these things?

When you're done with this analysis, you'll have created a true offer, by giving your products the adaptability of services, and your services the economics of products.

How to Manage an Offer

It bears repeating: If you think you are in a product business or a service business, think again. To survive and thrive, you must be in both. This probably means that your whole management approach (e.g., which variables you focus on, how you manage costs, your pricing philosophy) will have to change. The distinction between products and services runs deep in companies. Depending on whether you think of yourself as a product company or a service company, you think differently about all kinds of things. But dichotomies you thought always existed in product and service businesses will have to be resolved in light of the BLUR. The chart shows the ones that come immediately to mind for us, as explained in the text that follows.

MANAGEMENT MIND-SET

	Product	Service	Offer
Time Horizon	Time of Sale	Period of Contract	Life of Customer Need
Buyer "Care Abouts"	Price, Delivery, Convenience	Ongoing Support	Upgradeability
Dominant Cost Focus	Direct Costs	Period Costs	Design Costs
Source of Value	Manufacturing Process	Training, Maintenance	Platform
Design	Fixed, Uniform	Customized	Learning
Revenue Model	List Price	Subscription Period	Subscription & User Fees
Marketing Objective	Brand Loyalty	Relationship-building	Community-building

(CHARACTERISTICS)

TIME HORIZON

Product companies have always focused on a "moment of truth" in their business: the point at which the sale is made. Service businesses, on the other hand, have kept a more distant horizon in mind. They know they'll be dealing with a buyer over a period of time. Certainly, as products and services meld into offers, it's this longer time frame that will prevail. Indeed, the most successful offers will be those with the longest view—that focus on the total life cycle of a customer's need. In a limited sense, that's what GM founder Alfred Sloan had in mind when he created different lines of cars: He hoped to sell young men Chevrolets, then graduate them to Oldsmobiles as they became family heads, and finally, herd the fat cats among them into Cadillacs.

Today, offers can take advantage of Connectivity and upgradeability to serve customers' evolving needs. Let's say your customers are travelers. If you had a product mentality, you might produce a series of travel guides, and focus on bookstore sales. If you were service-oriented, you might run an agency that suggests itineraries and books reservations. But if you're offer-oriented, you might launch something like Epicurious.com, the Web site that employs all 10 attributes to create a virtual community for travelers. In addition to the guidebook content and the bookings services, the site gives people a forum in which to describe their past adventures and gather ideas for future ones. It serves their ongoing wanderlust.

Buyer "Care Abouts"

What do people care about most in choosing to buy? Managers in traditional product businesses tend to think it's price first and delivery and convenience second. Their counterparts in service businesses see it differently. To them, buyers care most about ongoing support. Again, managers of offers are in sympathy with the services people, but they push the idea of support even further. They believe that what matters to buyers is the ability to upgrade the offer as conditions change. It makes sense: If you could be guaranteed that the next computer you bought would get you through the next 10 years, wouldn't you pay a premium? Consumers are so intimidated by product obsolescence that they'll opt for low risk over low cost. Miele makes this explicit: It will download an upgrade to its dishwasher's wash cycles if detergents, dishes, or water quality change.

Dominant Cost Focus

Managers are paid to worry about costs. It's no good selling your offer hand over fist if your costs are eating up all your profits. So managers with a product mind-set focus on direct labor and raw materials—all those direct costs that go up with unit volume. Service-oriented managers, on the other hand, pay a lot more attention to period costs. Rent, payroll, and overhead are factors they expect to be their downfall.

Neither of these is the dominant cost focus of the offer manager. Instead, she sweats design costs. Think about software—and by extension, any software-intensive offer: The first copy costs a mint to produce; every subsequent copy costs pennies. Meanwhile, all that demand for upgradeability makes the upfront design even harder to achieve; it has to anticipate as accurately as possible where the business and technology environment is heading.

SOURCE OF VALUE

Another factor product and service managers consider differently is source of value; that is, the point in the business model where the most value is added. To product people, this lies in the manufacturing process. The product is as valuable as it's going to be when it rolls off the assembly line. To maximize this value, they focus on ideas such as "design for manufacture," something that ensures engineers will create designs that can be realized cost-efficiently. To service people, value is added downstream, in activities such as distribution, delivery, training, or providing the tools that facilitate maintenance. In a combined offer, tremendous value is added at many of these points, but the most value derives from the offer's underlying platform. This is what allows for future upgrades to be made efficiently, and thus is what the offer manager has to get right. Back to computers: If you can change the motherboard to upgrade the processor, add RAM capacity, buy a bigger hard drive, upgrade the modem and network cards, and connect to a new monitor, what did you buy? You bought into a platform that allowed you to upgrade as your needs and technology changed.

DESIGN

Perhaps the greatest dichotomy between product and service managers has been in their outlook on design. Products have always been characterized by their fixed and uniform design, whereas services have always featured customization. A carpenter is selling a service and not a product when he approaches a project with a clean slate. When he builds to stock, producing cabinet after cabinet of the same dimensions, he is selling a product. In the world of the offer, the design features customization, but goes a step further. The best offers are designed to *learn*, so that they evolve over time in line with their ongoing use. For managers of today's products and services, design for learning may be the most difficult shift in mind-set. But when you're following your customer for the lifetime of the need you're fulfilling, it can be the one with the greatest potential for payback.

REVENUE MODEL

What drives revenues? In product businesses, it's price, pure and simple. Setting it right is a managerial concern elevated to a science. Mark up goods and revenues will rise—unless they're marked up to the point that alienates customers. Finding the perfect per-unit price point is of paramount concern, and market researchers get paid a small fortune to study price elasticities of different market segments. In service businesses, managers have a different model. Their pricing is usually based on time periods and usage fees, such as per-hour charges for service people. But as offerers learn to observe their customers and their buying habits, the focus is shifting to the value of the customer over the lifetime of her requirement. The cellular phone business has analyzed this in detail. These companies know exactly how much profit each customer will bring over the next several years, and they refine their offer by providing free phone upgrades at customer-specific intervals to ensure they don't switch service providers.

So what's the right revenue model for an offer? One promising example is that employed by Claris Works' Users Group. Members pay a subscription price to join the group, and by virtue of membership, have the right to make purchases of software and services available only to the group. They also provide service indirectly by providing access to other users. The model combines subscriptions and user fees in a way many other offers could imitate.

Marketing Objective

When product and service businesses send their marketers into battle, what are their marching orders? In a product business, typically they're out to capture brand loyalty. The idea is to inspire repeat purchases, as a string of discrete transactions. Insofar as that is not assured by product quality alone, the weapon of choice is mass-media advertising. A brand a consumer has heard of is one she feels she can trust. Service businesses aim higher. They're attempting to establish ongoing relationships between themselves and their customers.

Relationship building on the golf course or in more substantive settings is the key to customer retention. Again, managers of an offer would do well not only to adopt the service mentality, but also to push it further. Some of the best offers are those that draw customers into relationships that include broader communities in which the customer wants to participate. Retention comes through a sense of belonging to a community brought together through the offer. For example, Procter & Gamble set up a community of interest on the Internet called Parentalk.com, a place where moms and dads can interact with each other and swap information on baby products.

And let's not forget the kids. Beanie Babies began as a product but is now also a Web site. Each "character" keeps a diary and corresponds with its fans. This community provides the company with information and also builds demand. The company "upgrades" by discontinuing popular models, which then become highly sought after.

Probably there are other aspects of management that will prove to be rooted in the world of either product or service that must be adapted to the blurred offer. We'd like to know what they are; and we'd like you to tell us. Remember, this book is not simply a book, it's an offer. By attaching it to a Web site—putting it online, making it interactive, allowing it to learn—we're attempting to practice what we preach. It isn't easy. We are professionals trained in a service industry, producing a thing—a book—that has always been sold as a product.

With BLUR, we're hoping to launch a successful offer. If we can do it, you can. But we can't do it without your help.

Have you considered the 10 attributes of offers? How can you apply them to your business? What changes in pricing, concept of value, management focus, and so on, would you have to make to manage your offer effectively?

And

Have you seen examples of these ideas? What questions do you have? Let us know at www.blursight.com.

CHAPTER 3

THE EXCHANGE

Every Buyer a Seller, Every Seller a Buyer

The difference between buyers and sellers blurs to the point where both are in a web of economic, information, and emotional exchange.

When Harley-Davidson customers buy their expensive, premium motorcycles, they're paying for much more than a high-flying hog. They are buying entry to a community of like-minded devotees who share a passion for all things Harley, be they branded clothing, decals, or even deodorants. The most committed bikers—whether they ride on the front of the seat or the back—further affirm their indelible loyalty to the brand with Harley-Davidson tattoos.

When Zagat's guides make money, collecting and publishing foodies' ratings of restaurants, the publisher is managing a multiple exchange: Customers get to compare notes and tell restaurant owners what they think of their offers, and they get a copy of the guide; restaurateurs get feedback on how to build business.

When Citibank provides private chat rooms on the Web for its customers, it enables them to get closer to each other—and to the bank. They can get advice on such topics as investing in real estate, or swap information with people of similar professional and financial goals. And Citicorp learns a lot about customer likes and dislikes.

Commerce used to be so simple. There were sellers and there were buyers. The seller brought a product or service to the table, and the buyer brought cash. The transaction was straightforward: The price was the price. Now, in an increasing number of business dealings, it's more difficult to determine just who is the buyer and who is the seller. A lot of the time, each is both. And even when those roles are clear, the form of payment is more convoluted. Parties are being compensated not just in money but in things like information and emotion. Thanks to the forces of BLUR—particularly the rise of Intangibles as a source of value and the spread of Connectivity—transactions are becoming anything but straightforward. The one thing you can be sure of is that the price is no longer the price.

A simple example is *slotting allowances*, the fees a consumer-goods manufacturer pays retailers for shelf space for a new product. For a foot of prime, eye-level position in the cookie aisle, you might pay up to $1,000 per shelf.[1] Think about this for a moment. The retailer is the customer who buys product from the manufacturer, the seller. So what's the seller doing giving money to the buyer? The manufacturer is buying marketing services from the retailer at the same time he's selling product to the retailer. They are engaging in an exchange.

At first glance, such payments seem like downright extortion. But they make good business sense when you realize that, until the product starts to sell, the retailer's real estate is more valuable than the product itself. The opportunity cost is high. The manufacturers know this, and they pony up the cash.

The distinction gets very hazy when you note that monetary payment is only a part of any given transaction. The real news in the BLUR economy is that other things—especially information and emotional engagement—make up a growing proportion of the value being exchanged in both directions. We have reached the point in our story where the Intangibles get serious.

Amazon.com, the online bookstore, is an elegant example. If the whole of that business were simply mail-order books, it would be dead in a week. Even with discounted prices to offset mailing costs, book buyers wouldn't tolerate the shipping delay. Amazon.com understands that its real edge over conventional bookstores is its capability to provide much more (and customized) information, collect and post reader recommendations, and create communities of like-minded people.

Customers pay Amazon.com not only in money but in various kinds of information that are valuable to its sales and marketing efforts. The Web site collects reader reviews and, each month, pays $1,000 in book vouchers to the amateur critic who submits the best-written item. The site also tracks virtual shopping baskets, so that it can inform someone looking at a title as to which other books were selected by fellow shoppers who bought that particular book. Meanwhile, thousands of visitors have alerted Amazon.com to their favorite authors, subjects, and book categories, and asked to be notified when new titles appear. That kind of self-selected, specialized marketing list is a direct marketer's dream.

Examples like this are going to become more the norm, which is why definitions of the terms "buyer" and "seller" aren't accurate anymore. They imply that the only exchange is the traditional, two-way affair where money is swapped for goods or services. The truth is, there are all kinds of value flying back and forth in a connected economy. And all these exchanges are happening so fast that there's no time—or need—to translate them into precise monetary terms. "Buyer" and "seller" just aren't descriptive enough of what's really going on. What we need to talk about instead is mutual exchange.

THE SIX-LANE EXCHANGE HIGHWAY

Seller

Emotion

Information

Economic

Economic

Information

Emotion

Buyer

The Six-Lane Highway

A simple metaphor for what's happening is to recognize how a two-lane road is being transformed into a six-lane highway. On yesterday's two-laner, only one lane of traffic went each way; the customer came from one direction with money while the seller came from the opposite one with a product or service. Now, we're talking about three lanes moving in each direction. As the illustration shows, one lane still conveys that economic traffic (albeit with a few new twists); the second is made up of informational exchanges; and the third carries exchanges of emotional value. These traffic patterns are key to the buyer/seller BLUR. They deserve a closer look.

THE ECONOMIC LANE

The traffic in the economic lanes looks familiar. It comprises the goods and the money to pay for them, as we've already discussed. But it's a more complicated flow because it includes variations that didn't exist before, new forms of economic value created by Connectivity, Speed, and Intangibles.

On the sellers' side, all kinds of new ways are being devised that enable them to provide economic benefit to buyers. Frequent flyer awards, of course, were a pioneering example, exchanging free travel for loyalty and customer information. Again, it's the opposite of the expected—sellers are supposed to get money, not give it. To many providers of goods and services, this feels like entering the economic highway from an off-ramp. But in a complex, highly integrated economy, there's no reason for money to flow in only one direction. Superquinn, a supermarket chain in Ireland, pays its customers to point out problems. If a chalkboard has a misspelling, the alert customer who spots it gets an apple pie as a reward. The customer who notices that the freezer thermostat is registering outside the correct range, or that a product on the shelf has passed its sell-by date, stands to get a cash reward.[2] This goes beyond self-service—the shopper is selling quality control services to the store.

What's happening when automakers offer rebates on new car sales? Basically, they're offering cash on the barrel for buyers to choose their model over another. On the manufacturers' bottom line, it nets out just like a sale price, but the mechanics of long-term financing give it a more attractive spin to the buyer. Typically, she will pay out the price of the car over time, but the rebate gives her a chunk of cash at the outset. Flip-flop that timing and you have the airlines' frequent flyer programs. Here, the carriers are exchanging transportation for money. How much money depends, in part, on the buyer's emotional connection to the seller (a frequent flyer upgrade, for example), as well as Intangibles such as choice. (Want to buy full flexibility? Pay full fare.) The accumulation of miles to redeem for free upgrades and round-trips is a form of payment to influence buying decisions. These programs have been so phenomenally popular in building repeat customers that they've spread quickly to every other kind of business imaginable, from supermarkets to shoeshine stands. What started out as schemes to BLUR the economic exchange through complicated discount structures became an emotional one. By treating its AAdvantage Platinum flyers specially, American Airlines creates a sense of satisfaction and membership that translates into loyalty.

Greater Connectivity is also causing a buyer/seller BLUR because it adds more people to a given exchange, in complex three-, four-, and more-way transactions. To understand how this works, consider first a business in which this has always been the case: magazine publishing. At first glance, it might seem that a publication such as *Vanity Fair* involves a clear buyer—the self-perceived "in" crowd—and a clear seller, the publisher. But think for a moment: Is the buyer here buying what the seller is selling? Actually not, because what the publisher, Condé Nast, is focused on selling is advertising (that is, reader's attention). At a subscription price of $12, magazine buyers barely scratch the surface of the costs of the editorial content they value. So it's the Armanis and the Breitlings of the world that are the real buyers of what readers are getting—and, of course, such advertisers are the most overt sellers in the magazine's pages.

The chain has other links, as well. Advertising agencies, representing the advertisers, prepare ads and may or may not be the same entities as the media buyers, who buy space from the magazine and earn a 15 percent commission on what an advertiser spends. The publisher adds to its revenues by selling its subscriber mailing list to other publications and direct marketers of products and services. So the popular editorial contributors are making money for the media buyers by increasing circulation, and for the publisher by enriching the mailing list.

This magazine model is, of course, the same one used by television and radio. Even telephone companies are starting to experiment with it. In Paris today, residents can choose to have local phone service provided at no financial cost, provided they are willing to listen

to a 15-second ad before they make each phone call. And because
the advertiser-medium-subscriber triangle is so well understood, it's
spilling over into other media, like the World Wide Web. For many
Web-based businesses, including PointCast (the customized news service we've already talked about), the sale of on-screen advertising is
the only source of revenue. The advertisers wholly subsidize the services that users get free of charge.

At least one Web-based business, Juno, has pushed things a little
further, adopting a business model that resembles a media buyer more
than a magazine. It obtains the right to show advertising messages to
subscribers (in essence, buying ad space on their screens) by giving
them something they value in return: free e-mail service and software. It's a perfect example of a BLUR-era exchange. Is Juno the buyer
here, the one paying for the software and services? Or is the user the
buyer, since he is the one paying attention? Or is it the advertiser,
who really wants the attention?

Systems of multiple exchanges like this are more frequently
replacing traditional one-seller-to-one-buyer transactions. And clearly,
the magazine model is not the only way in which multiple buyers and
multiple sellers can be brought together in complex, multidirectional
economic exchanges. Where is American Airlines making its profit?
In fees charged for "publishing" ads on the in-flight movie screen?
From commissions on Airfone calls? Or consider the supermarket,
which may make more in advertising—the slotting allowance—than
on the groceries. All manner of other enriching possibilities are opening up. It won't be long, for example, before we see groups of con-

sumers—holders of the American Airlines credit card, perhaps—joining up online to buy blocks of tickets with the power to drive great price deals. Members wouldn't need to be in the same neighborhood, city, or state. Similarly, multiple producers could join up to create bundled offerings that fetch premium prices on the market. An example might be a car, for which the monthly payments cover debt service on the purchase, insurance, fuel, and scheduled maintenance. Or an off-the-rack suit, bundled with accessories along with vouchers for dry cleaning. We can even see local utilities getting into this—or distant ones, for that matter, once that industry is deregulated. Why not buy white goods and the juice to run them in one contract? For a family with a lot of laundry—frequent dryers—the overall price could drop with usage. Better yet, why couldn't groups of households equipped with solar panels channel their excess power to utilities in exchange for cable television service? (Again, who's buying? Who's selling?)

With sufficient Connectivity in the economy, there's no logical or practical limit to how many buyers and sellers can be in an exchange—or how many can participate on both sides. Ultimately, in the economic lane, a universal, chip-driven "cash register" may simply collect tolls from buyers, make special price adjustments for volume and regular travel, and send payments to various sellers, many of whom are far removed from an initial point of sale. Doesn't this describe the movie industry today? The star gets a percentage of the gross, the cinematographer gets a negotiated fee, Technicolor gets a per-print price, and the price a moviegoer pays for a ticket depends on the time of the show and the customer's age. The point is, every business will operate this way. The net result is that offers are being exchanged among a web of players, not just for money, but for other Intangible value as well.

THE INFORMATION LANE

Where exchanges of value are really starting to get interesting—and complicated—is in the middle lanes of the six-lane highway. These carry a different kind of value than the purely economic: They support information exchanges. Again, this is not a one-way street. At least as much information of value is flowing from buyers to sellers as in the opposite direction.

This wasn't always the case. It used to be that buyers were fairly mute in a transaction. The information that was changing hands was going mainly from seller to buyer, and took very conventional forms. The seller of a new appliance would provide an owner's manual. Or Clairol, for example, provided an 800-number that you could call for help should your hair turn the wrong color. But where it was provided, the information value was an afterthought, a little bonus to an essentially economic transaction. We've come a long way from those days. Now providers of goods and services are seeing the information content of their offering as the highest source of value-added, and the driver of higher profit margins. At the same time, they see the growing value of information that comes from buyers. If anything, they're more eager to listen than to inform, and they are constructing elaborate mechanisms to hear the voice of the market.

American Airlines' Eaasy Sabre system is now regarded as the pioneer of an information channel. Nevertheless, American has con-

tinued to expand its bandwidth. Now that flyers can connect to
Eaasy Sabre online, the possibilities have expanded again. The company can monitor its customers choosing flights and fares, enabling it
to make them special offers on the spot based on a specific flight's
load factor; test price elasticities; advertise tie-ins in the destination
city likely to be of interest—all well beyond the information traffic a
travel agent could bear.

Another good example is Peapod, the online grocery shopping
and delivery service. Peapod brings back the days when folks could
call in their grocery orders and have delivery to their doorstep a few
hours later—eggs uncracked, ice cream unmelted. Here's how it
works: The shopper logs on to Peapod's system via a personal computer and modem, selects purchases from lists of available items, and
gives a 90-minute period when she wants delivery. Peapod dispatches
the order to an employee standing by in a nearby grocery store. The
employee picks the items from the shelves, takes the order through
the checkout, and hands it over to a Peapod driver. In its first four
years of operation, Peapod built a combined customer base of 7,500
in Chicago and San Francisco, with a retention rate of 80 percent.
Its volume in those cities accounts for 15 percent of total sales in
the 12 Jewel and Safeway supermarkets where its employees do the
shopping.

The interesting question is why the service is proving so attractive. It's far from free; users pay $29.95 up front for application soft-
ware to access Peapod's database, plus a monthly service charge of
$4.95. Once they're in the system, they pay another $4.95 for each

delivery, plus a 5 percent surcharge on the total order. Even though this sounds expensive, Peapod clearly hasn't yet hit the limit to how much people will pay to save a trip to the store. In fact, most customers report that, despite the hefty charges, using Peapod actually reduces their grocery bills.

The savings stem from the value of the information Peapod provides. Shoppers can view product listings by any number of criteria (item, category, brand, or special sale), and then sort the list according to what's most important to them. Usually, this is unit price, although it can just as easily be calorie content. Let's say you want a can of tuna. The database instantly displays your options: water-packed versus oil-packed, 5 oz. versus 8 oz. versus 12 oz., Star-Kist versus Bumble Bee, and so on. A quick sort shows you the best price per ounce. Peapod customers comparison-shop more assiduously, and buy fewer impulse items. And since they're sitting at a desk and not strolling around a store, they're more likely to use coupons. (Yes, Peapod honors them.) If the store is out of what is ordered, a Peapod employee will call the customer and ask about substitution. All in all, quite a value proposition.

Skyway Freight Systems is another company that understands information is the greatest value it brings to exchanges. A commercial shipping subsidiary of Union Pacific, Skyway realized early that a major source of differentiation in its business would be to give customers real-time access to the status of their shipments, across all modes and territories. To do this, it launched the industry's first electronic packing slip and a satellite tracking system. Once that kind of Connectivity and visibility was in place, it became clear to Skyway that it could play a greater role in coordinating activities on either end of its service: the manufacturers' efforts to produce the right products to meet demands and the retailers' efforts to keep items in stock. Its innovative approach to that intermediary role has been to create a system called Concerto. Concerto takes actual sales information from retailers and uses it to drive the entire process of making the right product and getting it into consumers' hands, from initial sourcing of raw materials through delivery to the retail floor or the consumer's door. By doing so, it cuts supply chain costs for everyone, while boosting revenues by accelerating products' Speed to market. Note, too, that in true BLUR fashion, Concerto creates exchanges that draw in many buyers and sellers, including competing suppliers, whose inventories are alternately consolidated, separated, and diverted as needed.

Other companies are finding that customers value the information component of their business most. Consequently, the information isn't just an afterthought to economic activity anymore: It's the tail that's wagging the dog. And just as producers of goods and services come to realize that the information they offer is valuable, so will consumers. In the past, current or potential buyers were just hoping to be asked for their opinion, but not too much longer. Consumers understand now that their feedback has real value, and that if they take an hour filling out a survey, for example, they know that the information they supply has value in the marketplace. So they're saying, "Pay me." And they're not looking for a one-dollar coupon. Previously it was possible to convene a focus group to react to a new product just by issuing invitations and sandwiches. People were flattered they were asked, and eager to see the newest thing. But they're over that now. To attract a quorum today for a focus group, you might have to offer each participant upward of $100 an hour.

Nevertheless, smart sellers are happy to pay cash for the information customers can provide, whether it's what they spend on furniture, where they go on vacation, what their hobbies are, the sports equipment they buy, the computer software they've installed, or how much gasoline they use. The problem is, in most cases, no convenient mechanism exists for the exchange of feedback for greenbacks. Surveys and focus groups are one-time events: How do you reward people who provide useful information on a continuous and incremental basis? For now, most companies try to respond in kind. They may give product for information. A consumer with a useful comment on PopTarts flavors might, for example, get a coupon toward a future purchase. The St. Francis Hotel in Santa Fe, New Mexico, mails a 20 percent discount certificate to the customers who fill out the comment card. More commonly today, companies try to reward information with information. The growth of corporate Web sites is part of this motivation; they are a source of information from the firm to the public, and a channel through which the public can volunteer thoughts to the firm.

One company that has made information exchange an explicit part of its offer is Web-based Mainspring, Inc., which runs online conferences on electronic commerce and sells access to its services and rich store of case studies. Corporate clients usually sign up because they are undertaking electronic commerce ventures themselves and want the benefit of others' experience. But it's not a cheap club; the basic annual membership rate is $20,000.

But here's the interesting part: Mainspring recognizes that it is selling to precisely the population that represents the greatest potential source of new material. As its members make progress in their own efforts, Mainspring would like to study them, write up case studies, and enrich its knowledge-based offering even further. It's more than willing to strike a deal for this input. If you agree to be a case study, your membership drops to $15,000. The balance of your dues takes the form of the information you provide. Mainspring sends a team to put together a story on your goals, methods, accomplishments, all of which becomes available to fellow members. The buyer pays less and gets the benefit of the study as well.

Unfortunately, not all information seekers are as willing to pay for the value they receive. The infamous recent example of this has been the abuse of "cookies"[3] on World Wide Web sites. Cookies were created for a useful purpose by Web browser companies such as Netscape to capture preferences of individual users and store them on those users' hard drives. This way, every time a user pulls up a page, it reflects any customization he has performed on it in the past. But because cookies collect their information surreptitiously, they have proven useful in other ways. Every time you visit a cookie page, you leave a calling card that reveals the kind of computer you have, your network location, and many other details. To the outrage of customers, Microsoft's cookies in Internet Explorer even inventoried other software a new subscriber had installed. What's more, if you voluntarily "register" at a site by giving it your name and address, all this information can be recorded and linked directly to you. That information, in turn, can be sold to others, such as consumer marketing organizations.

Not surprisingly, buyer resistance is growing to anything being handed out without their express permission. In the case of Microsoft, buyer pressure convinced management to toss its cookies—at least the particularly invasive ones. Again, we're seeing the effect of a new breed of buyers who know the value of their information and are prepared to collect on it. The concern for "privacy" is often a misnomer for what customers rightly see as a theft of value.

What's next? As more consumers begin to recognize both the value of their information, and their ability to connect and pool that value, we may see:

- Organized groups of customers who collect information about themselves and offer it for a price.
- Clusters of customers focused on a particular product or service and telling the marketplace what they want.
- Customers of a particular enterprise mobilizing to upgrade and improve current offers and to demand new ones.

Electronic directories of such groups will not be far behind. An individual can find the buyer group of choice and join up, not to give away information, but to offer it in return for something of value. The information lanes are carrying heavier traffic by the day as

Intangibles and Connectivity grow. The volume already rivals the economic aspects of the exchange. Now, let's look at the construction project underway to expand the highway.

EMOTIONAL EXCHANGES

The last pair of lanes on our six-lane highway is a little less busy but is definitely picking up Speed. These lanes convey all the emotional value that goes back and forth among providers and consumers of goods and services. Like information, this kind of value has always been around, but generally subjugated to the economic transaction. What we're seeing now is a more conscious recognition of real value in these lanes with price tags (monetary and nonmonetary) placed on Intangibles such as loyalty, esteem, excitement, and engagement.

Our favorite example is Harley-Davidson. As our chapter opener suggests, customers who buy a Harley feel they're getting much more than a motorcycle; they're buying into a lifestyle, an attitude, an image. Think about the term "biker." Does it summon the image of a Honda? A Ducati, perhaps? No way. If you want that persona, you gotta have a hog. "There's a high degree of emotion that drives our success," Harley-Davidson CEO Richard Teerlink told *Fortune*. "We symbolize the feelings of freedom and independence that people really want in this stressful world."[4] Harley's profit margins are higher than many of its competitors'; the premium it earns is for the emotional value it brings to the transaction. And naturally, it's repaid in emotion, as well. The degree of loyalty and pride of ownership that

Harley riders exhibit is truly phenomenal. How many other products could inspire tough guys to wear conspicuously labeled merchandise? Harley is besieged by manufacturers seeking licenses to make all manner of branded clothing and accessories. (Okay, so many of the tattooed bikers you see at rest stops are examples of copyright infringement. *You* tell 'em!)

What's good for motorcycles is good for all offers in the marketplace. Nike knows that and focuses its substantial advertising investments on eliciting emotion rather than showcasing product features or even selling benefits. MetLife knows it and uses the *Peanuts* characters to put a friendly face on a business rooted in mortality tables and actuarial science. Even the statistics-obsessed mutual fund industry has gotten the message. Scudder, Stevens & Clark turned to emotion-triggering advertisements, such as an investor discussing her children's savings with a Scudder telephone representative. "The most successful ads are relationship-oriented and not hard-driving product ads," reports Scudder chairman Mark Casady. On days when such commercials run, the company receives hundreds of additional calls from investors.

But emotional value from companies isn't just the result of advertising—good vibes around a brand lead to loyal repeat purchases. Companies are learning to engage customers on other levels in a true give-and-take of emotional value. The Big Boy chain of diner-style restaurants did this by sharing its own mixed feelings about its mascot. "Big Boy: Should he stay or should he go?" was the question it posed to the BLT-eating public. In the end, he stayed. It wasn't all a gimmick, though. Management truly had not decided; and it valued the public's emotions for Big Boy more than it trusted its own.

MCI, as we all learned during the dinner hour, tried to win the battle of the phone companies by associating its service with friends and family. On one level, it was simply using the old tactic of rewarding existing customers for successful referrals. But the potential was for a much deeper engagement; if a caller came to think of MCI as "the service all my friends use," her regard for those friends might reasonably be extended to the firm. The promise of emotional value wasn't hollow on MCI's part, by the way. By discounting all the calls placed within a person's network, it effectively invested in the emotional lives of its customers over the life of that ad campaign.

Affinity, as in the special-interest credit cards being offered left and right, is a fertile field for emotion-based businesses. In many cases, the point of the card is to channel some percentage of the profits that would normally accrue to the issuer to a charity or nonprofit group valued by the cardholder. One example is the card offered by the Aircraft Owners and Pilots Association, AOPA. Using the card is a painless way to donate to AOPA's lobbying and education efforts to support private aviation. If you love AOPA, you'll love this card. And you may even extend that love to its issuer, MBNA America, the institution that issues the card. After all, it clearly shares your view of AOPA as a worthy cause.

As with so many things, it may be the Internet that really enables emotional value to come into its own as a currency of exchange. Discussion forums can get pretty ebullient—and, by turns, pretty vitriolic. Companies that build forums into their Web sites may have a greater capability to tap into the power and value of all that emotion. Already, the Internet is proving a fertile ground for building rich customer relationships. Its power to engage customers anywhere and anyplace was dramatized in the summer of '97 by a "Dear Amazon.com customer" letter e-mailed by the Internet bookstore. It invited subscribers to help author John Updike write a short story, online, over the course of 44 days. Each day, Amazon editors chose from thousands of paragraphs submitted by customers as possible ways to advance the story line. Every day, one talented writer would collect $1,000 for the chosen submission. At the end, one of those contestants got $100,000. To Updike fell the task of devising, and writing, the ending.

Of course, what the bookstore got in exchange was thousands of customers stopping by to browse its offerings, 44 days in a row. The buyer/seller lesson? The Updike caper showed a terrific relationship at work: enthusiastic buyers, who were happy to invest time and presence, and a seller prepared to meet them more than halfway, not only with cash but the *frisson* of collaboration with a world-renowned author. All in all, a happy ending.[5]

Buyer Redux

A common theme emerges from this discussion of exchange: The traditional buyer in the equation is gaining power and leverage relative to the seller. In the industrial model, the economic benefits of mass production created a one-way relationship, in which the manufactur-er defined the product, set the price, and established the time and place of purchase. The buyer was price taker, as economists would say, and the seller a price maker. This imbalance is being redressed. After all, it wasn't always this way; the local grocer used to tell you what was good today (information) and ask about your kids (emo-tion). And you might point out a spot or a bruise on the melon and get a better price (economic). But now, these relationships needn't be local, or quite as simple.

In all markets, rather than simply being expected to fork over cash for an offer, the user is in a position to get economic benefits from a two-way, multifaceted exchange. Rather than just being a pas-sive recipient of sporadic information, he is seen as a source of valu-able insight and opinion. And instead of being a sap to be manipulat-ed by emotionally charged advertising messages, he is actively engaged in an exchange of pride, satisfaction, and loyalty.

When this transformation is complete, the traditional transaction will be converted into an exchange, blurring the roles of all parties. The table below identifies some of the dimensions of this transforma-tion. In Part III, The BLUR of Fulfillment, we'll cover some of its most startling implications.

The Value 500

As power continues to shift from producers to consumers, we predict that a new organizational power elite will emerge: collections of con-

CHARACTERISTICS OF THE EXCHANGE

	Traditional Transactions		BLUR Transactions
	Seller	Buyer	Exchange
Value Role	Create	Consume	Both Create and Consume
Value Received	Money	Product or Service Utility	Economic, Informational, and Emotional Value
Communication Role	Sender	Receiver	Interacter
Information Role	In Control	Limited Access	Shared Access and Creation
Relevant Time	Business Hours	Business Hours	Continual and Connected
Relevant Space	Point of Sale	Marketplace	Connected Anywhere

Characteristics

sumers with enough combined clout to influence economic direction. The group that has historically held that powerful position is, of course, the Fortune 500. It's interesting to reflect that, for most of its existence, *Fortune's* annual listing of the largest U.S.-based corporations was focused exclusively on industrial manufacturers, as measured by their revenues. More recently, service companies were given their own list. As companies evolved, adding services to a product mix, the lists became confusing. General Electric, for example, migrated from the Industrial 500 to the Service 500 as its GE Capital and other financial service operations became a dominant source of revenues. Only in the mid-1990s did the magazine combine the two lists. Even in its new form, however, the Fortune 500 is still all about concentration of power among the nation's largest *providers*.

Over that same time period, power within those organizations has migrated slowly but inevitably toward the periphery; customer-facing units have become the movers and shakers, manufacturing has lost much of its clout, and headquarter functions have atrophied. The direction of the power shift is clear: It went toward the customer. And ultimately, it will shift right off the premises and squarely into the hands of the customers.

Enter the Internet. With the unprecedented level of Connec-tivity it grants individuals, consumers now have the means to com-bine that power into enormously powerful blocks. And don't think they won't be inclined to do so. A student in Iowa who wanted to initiate a demonstration around a political issue broadcast an e-mail that brought thousands together in 48 hours. Before long, there will be enough organizations of consumers wielding enough market sway that it will be time for somebody to create a new list that will reflect the accumulation of desires, not sales. It will be about delivering the greatest value to consumers. We'd call it the Value 500.

Here's what a Value 500 company might be like. Take a look at American Association of Retired Persons (AARP). AARP is the political lobby to end all lobbies, but it's also an economic lobby. It has turned its loyal and like-minded membership into a significant and powerful market force, negotiating member rates for all manner of services and products. Value 500 members will also have features in common with entities such as Gartner Group and J.D. Powers, which unite the voices of customers and turn them into another chorus to be reckoned with. The affiliation marketing business, which already provides an affinity card for every conceivable micro-segment, will spawn the aggregation of consumer power. Customer institutions will achieve the scale of provider institutions. Ready for General Airline Passenger as Number 1 on the Value 500 list?

Perhaps we'll all be reading *Modern Maturity* before this prediction comes fully to pass, but the course is already set. In an economy where buyers and sellers are blurring, and one-dimensional transactions are giving way to rich, multidimensional exchanges, it makes all the sense in the world. Exchanges take all forms—economic, informational, and emotional—but the bottom line has to be value.

An historical note: The last such power shift was the creation of organized labor. The AFL-CIO, Teamsters, IBEW, and the rest were the "Wages 500" of their day, a mechanism to aggregate the power of labor to deal with the industrial behemoths. In that tangible world, heads were cracked—and lives were lost—in an attempt to block any shift of power away from the corporation. This time the coup will be bloodless, propelled by the opportunity to serve customers better and accomplished by the two-way exchange of value over the connected infrastructure.

Is your business prepared to organize its customers or are

you going to wait for "outside agitators" to arrive?

When the members of the Value 500 take their place on the economic stage, the nature of desires will have fully blurred. From discrete sales of products at arm's length to continual exchanges of value co-created by economic agents, each providing and consuming value simultaneously, we will be well into the development of a new economic structure. This won't be a value chain, a linear connection from seller to buyer (note the industrial directionality), but a complex set of exchanged offers. This is the economic web, the subject of the next chapter, and a radical change in how the economy achieves the fulfillment of desires.

Can you identify the informational and emotional compo-

nents of your exchanges, in both buyer and seller directions

simultaneously? With both customers and suppliers? What

value are you, or they, leaving on the table? Which elements

could be repackaged and exchanged with others? How could

your value exchange be more continual and interactive, any-

time, anyplace? How would you treat your customers differ-

ently if you thought of them as an organized group?

Fulfillment

In our market economies, we have grown accustomed to the idea of corporate strategy representing the design of "what" our businesses do, and organization representing the "how." Strategy obeys the laws of economics; organizations are governed largely by the laws of power, status, and psychology. As Connectivity blurs the boundary of the enterprise, this distinction will disappear, and economic signals increasingly will be used more often to drive agility and adaptation.

In Part III, we discard the zero-sum, efficiency-driven concept of competitive strategy that has dominated the mature industrial economy of the past 50 years in favor of a set of ideas about economic webs, designed for innovation, adaptation, and growth. Since organizations face the same pressures for agility and growth as does the economy as a whole, they will turn into organizational webs, running by the same rules as the economic webs.

PART III

The
BLUR
of
Fulfillment

CHAPTER 4

THE ECONOMIC
WEB

Manage the Supermarket Like the Stock Market

The BLUR of businesses has created a new economic model in which returns increase rather
than diminish; supermarkets mimic stock markets;
and you want the market—not your strategy—to
price, market, and manage your offer.

Netscape, convinced that critical mass was essential if its Navigator Web browser software was to succeed in the highly competitive marketplace, made an extraordinary move: It gave away 40 million copies of its software. Did the move devalue the product? No way. By establishing huge market share fast, it effectively positioned Navigator as the standard—and made it extremely valuable to Netscape's content providers and advertisers.

Sun Microsystems executed a similar move, but took it even further. It gave its Java programming language away to everyone. What's more, it formed a venture capital fund to nurture start-ups developing Java applications. The idea: By establishing Java as the standard, Sun's potential to sell servers goes through the roof.

In the market for a car? If you already know what you want, Auto-by-Tel will do the haggling for you. The Internet-based company links with multiple dealers in a region for every kind of car and puts customer orders "out to bid." The service is a boon to consumers, and it overturns the economics of car sales. What dealers lose in markup they save in advertising. Hardly an obscure start-up, Auto-by-Tel advertised during the 1998 Super Bowl broadcast, the most expensive time slot on television.

Traditionally, a wildly successful movie (or, let's go back even further, a hit radio show) would spawn a slew of merchandising deals. Today, Disney makes such exchanges an integral part of its rollout. All that product and co-promotion creates ambient advertising that feeds back directly into additional theater and video sales.

In a business world of increasing Speed, value-driving Intangibles, and pervasive Connectivity, the eyes of corporate strategists are blurring in unsettling ways. The focus of strategy has always been to position a company in relation to its business environment. It against the world. But what happens when the distinction between the company and its environment is blurred? When it's no longer clear where the company leaves off and the world begins? When the fates of diverse companies are linked like rock climbers on a sheer cliff? When even a competitor's well-being can be vital to your success?

If the boundaries of strategy are blurring, then the way to arrive at a good one is also unclear. The fundamentals that have always guided strategic planning are shifting. All that we practice about macro- and microeconomics was formulated to reflect an industrial economy. But that was a world that was much slower, rooted in physical goods, and marked by the actions of isolated actors—in a word, unblurred—so that it behaved in utterly different ways. Your college economics theories just don't describe the real world today. Their assumptions are mostly invalid, and many of their deeply held tenets—the relationship of inflation and unemployment, or deficits and interest rates, or trade gaps and currency values, or book value and market capitalization—are living on their reputations, not on relevant empirical evidence.

It's no surprise that, for some people, the BLUR seems to preclude all possibility of a strategy.[1] Why pretend we can make our own future when it's obviously out of our hands? What chance is there to predict accurately when the laws of the new economy aren't known?

What good are plans if the Speed of business today means they are obsolete before an organization can understand and implement them?

But before you lay off your strategists, consider that there are still decisions to be made about the company in relation to its environment. If it's true that you're more dependent on outsiders to succeed, you need a way of spotting the right players and creating the right relationships. If prediction is futile, you need ways to build in more flexibility, to catch the curls as they come. If long-term plans are useless, good short-term ones are more valuable than ever. As you give in to BLUR, you'll find that new rules of economic behavior are yielding new rules of business strategy.

Who's in Your Web? How Do They Relate? How Do They Add Value?

The truth is, no company today can act alone. Success arises from networks—or as Harvard Business School Professor Adam Brandenburger says, "economic webs," in which everyone is an active player in creating and claiming value."[2] Though such networks can be electronically enabled, the economic web is essentially all about relationships. And BLUR's trinity—Connectivity, Speed, and Intangibles—is the force that's making these relationships so unavoidable, or better, *right*.

The need for Speed means that many companies that want new capabilities don't have time to grow their own. On the other hand, the probability of further rapid change makes out-and-out acquisitions too risky, while the ease of electronically connected coordination makes them unnecessary. Presto! A new density of connections among companies, industries, and individuals. The Intangibles are the value of all these new relations, some of which are downright surprising given their surface incompatibility.

The ideal economic web is a constellation of players coalescing quickly around an emerging business opportunity, and dissipating just as rapidly once it runs its course. An economy driven by Intangibles favors economic webs, because nongoods like knowledge and information are the lingua franca of networks. And high levels of Connectivity mean that economic webs can be established more quickly and operate more effectively, because they are more tightly knit. Instant interactivity makes many more webs possible. Strike that: It makes them inevitable.

Of course, some opportunities, such as the "Wintel" virtual alliance between Intel and Microsoft, may last quite a while. In those cases, relationships must be chosen not only for the response they enable right now, but for what they offer long term. Some would argue that the choice of such "generative relationships" has become the only strategic decision.[3]

Economic webs have always existed in elementary form. They spring up wherever one company's well-being hinges on the success of others. In the past, this was usually the result of geography. Think of a shopping mall. The Sharper Image would seem to have nothing in common with Talbot's next door, but each depends heavily on the traffic generated by its neighbor. Colocation also means shared labor pools. Companies starting up today in Silicon Valley or in North Carolina's Research Triangle Park benefit from the critical mass of local talent. Critical mass within a region also boosts market awareness. Dozens of biotech start-ups got more attention—and undoubtedly more capital—when they were clumped together and called "the Massachusetts Miracle." The same phenomenon often happens in the arts. Impressionists made a splash because there was a school of them. Similarly, thanks to a thriving Chicago theater scene, even small repertory companies can be profitable in the Windy City. And a band like Pearl Jam might still be playing dive bars if it weren't part of the "Seattle sound."

What's happening now is that economic webs are extending far beyond these kinds of local interactions and dependencies. Global Connectivity is stretching the linkage to the far corners of the economic world. Companies can be oceans apart and still climb into the same boat.

Take the economic web spun by the U.S. apparel industry in the eighties—the "quick response" system, as it came to be known. This is an industry in which the players' fortunes are clearly interwoven. With the changing tastes of fashion setting the pace, profits depend on tight coordination among fabric makers, clothing manufacturers, transportation providers, and retailers—all on different continents. Benetton gained fame more than a decade ago for engineering a system so seamless it cut months out of the traditional supply chain. And because the company could tie its production to retail activity, it kept the hottest items in stock and was left with little to unload in end-of-season sales. Speed was the driver. Intangibles—the vagaries of fashion, the ideas of designers, the end-of-day sales records for each store and stock keeping units—were the keys. Connectivity is what enabled the system to deliver.

As the pioneer, Benetton built its own web and proved that the profit potential in such quick response is compelling. But to pursue that strategy today requires what many perceive as too big a risk. For one thing, you have to place your fate more squarely in the hands of companies and management teams outside your jurisdiction. In other words, you must give up the idea of control and rely instead on your ability to influence. This is the challenge of membership in any economic web. More important, this is the wrong risk to worry about. Instead, you should fear being late to market with capabilities that are less than world class. Besides, why would an outside company whose livelihood depends on your satisfaction not do as good a job as a group within your organization? The ability to recognize this imperative—not to outsource, per se, but to establish an economic web where you can thrive independently—is one of the keys to a successful BLUR strategy.

Of course, this challenge can be intimidating, particularly when the web involved isn't so straightforward as a supply chain. After all, it isn't quite so difficult to accept that one's fortunes are linked to suppliers upstream and distributors downstream, a concept that has been around for a while. It informed the Japanese keiretsu formed by "channel captains" such as NEC, Canon, and Nikon and, in a less one-sided version, was critical to just-in-time inventory management, which Toyota pioneered in the seventies. It moved even closer to mutual benefit when Procter & Gamble decided to play in the same sandbox as Wal-Mart in the diaper business. Wal-Mart kept P&G up to date on what was selling in which stores; in return, P&G was able to stay on top of Wal-Mart's inventory needs. Revolutionary at the time, customers and suppliers acting as partners like this is now commonplace.

Economic webs can become threatening when they start embroiling you in the affairs of competitors—or perhaps even worse, of companies you don't know much about at all. When IBM and Apple teamed up to create the PowerPC, it wasn't because they had decided to stop competing. But each was willing to bet that greater Speed in technology development would more than offset the risks of their temporarily joining forces. The IBM-Apple deal was by no means this simple, however. Their third partner was Motorola, maker of the 680x0 microprocessor on which Apple's Mac operating system ran. Motorola competes with Intel, which makes chips for IBM's PCs. An added complexity was that Claris, a maker of application software for both Mac and Windows, is a subsidiary of Apple. Sounds a bit like the Hatfields and the McCoys deciding to hold a joint family reunion, doesn't it?

Intel also figures as the exemplar in the realm of "co-opetition."[4] Intel not only makes ever faster computer chips, but also tries to ensure that there's a growing market for them. One effort the company makes is to work actively with MCI to advance the development of greater communications bandwidth. Once phone lines can carry more data, demand will rise for more powerful processors to bring it effectively to desktops. Again, the basic idea is not unprecedented. Automakers understood early that they would sell more cars if they could help certain key "complementors" such as car finance companies and repair businesses. Michelin knew long ago that travel guides would promote tourism and the use of tires in the process. But the potential for these interdependencies, and for aiding and abetting complementary businesses, grows exponentially in an economy that is connected and driven by Intangibles.

By the way, all that work by Intel should make their complementors, who assemble Pentium machines—IBM, Compaq, et al.— happy, right? Well, yes, except for that niggling fear that Intel will start producing its own competitive machines. Why else establish a consumer brand—Intel Inside—for a piece of silicon, they wonder? Sometimes your neighbor in the web *is* going to take on the look of a predatory spider. Is Intel cooperating by building the market—or competing by building its brand? Both.

A perfect example is what's going on now in the computer game industry. As Kevin Kelly, editor of *Wired* and author of the powerful *Out of Control* notes, "Game companies . . . devote as much energy promoting the platform—the tangle of users, developers, hardware manufacturers, etc.—as they do to their product. Unless their web thrives, they die."[5] It's the old VHS versus Beta battle playing out all over again. If you're Sega, you can simply sit by, churning out your own product and hoping the industry shakes out to your advantage. Or you can get your assets in gear and make it happen.

That's certainly what Sun Microsystems is trying to do with Java. As touched on earlier, in one of the more innovative approaches to weaving of economic webs, Sun joined with eight other interested parties to create a $100 million venture capital fund specializing in Java-related start-ups. The partners turned to Kleiner, Perkins, Caufield & Byers, the preeminent venture capitalist firm of Silicon Valley, to manage the fund. Consider that this is money that could have gone toward product development, to marketing Java directly, or more to the point, toward selling servers. Sun, however, is convinced that the most important driver of its success is an industry awash in Java. Every Java developer that succeeds, this thinking goes, contributes to the demand for Sun's SPARC stations. Every one that fails casts a shadow across Sun's future. This understanding of Web-based competition is behind a Sun suit against Microsoft. It alleges, essentially, that by creating a noncompatible version of Java, Microsoft has poured salt into the coffee. Java's chief value is that it initiates Connectivity for all its users. A disconnected version renders Java nearly worthless.

About four centuries ago, John Donne put it this way: "No man is an island, entire of itself; every man is a piece of the continent, a part of the main. If a clod be washed away by the sea, Europe is the less . . . any man's death diminishes me, because I am involved in mankind; and therefore never send to know for whom the bell tolls; it tolls for thee."[6]

Kelly recently wrote the strategic version of the same thing (albeit in *Wired* speak): "There is no future for hermetically sealed closed systems in the network economy. The more dimensions accessible to member input and creation, the more increasing returns can animate the network, the more the system will feed on itself and prosper. The less it allows of these, the more it will be bypassed."[7]

So good luck finding the point where you stop and the market begins. The distinction between a company and the environment it exists in is becoming hazier. Your agenda is all tangled up with mine. It's a BLUR.

The More You Use, the More You Have

When Kelly talks about "increasing returns," he's referring in large part to the work of economist Brian Arthur, who has done more than anyone to rewrite economics for the postindustrial age. Arthur's observation is that, in an information age marked by Connectivity, Speed, and Intangibles, a curious thing happens to the classic pattern of diminishing returns. It doesn't happen.

At first, this sounds like heresy. Diminishing returns is a pillar of nineteenth-century economics, after all. The core idea is that there are economies of scale in production—purchasing power, three-shift operation, energy efficiency, and so forth—but that these improvements become proportionally less important as size grows. They get swamped by costs of coordination, which include those big headquarters organizations of the past, and by transportation, which might mean moving all the output of your giant steel mill to distant markets.

This explains why economists, known for rarely agreeing about anything, tend to subscribe unanimously to the belief that "nothing grows
to the sky." In other words, no company satisfies all the demand,
because multiple firms can do so much more efficiently. In this view, if
one company does dominate, it can only be from exercising the type
of power that stems from monopoly. Indeed, this is the basis of the
antitrust laws developed in the 1890s.

Let's BLUR that. What if connection makes coordination easy and
cheap (economic webs)? What if Intangibles don't cost anything to
transport and can reach customers quickly? Then those costs won't
rise as you get bigger. In such a business—software particularly—most
of the cost is in the development of the product, and there are compelling economics to spread them over the largest number of copies.
Direct costs are dead, and diminishing returns died with them, a vic-
tim of Intangibles.

But there's an even more powerful force driving increasing
returns: Connectivity. Imagine that you make a product in which
each unit sold has greater value than the previous one. In other
words, you could charge more for the ten-thousandth unit than you
did for the tenth. Sound absurd? How much would you pay for the
world's only fax machine? It would be useless. But the more people
who have them, the more use you'd get out of yours—and the more
you'd value it. *Similarly, the value to the web as a whole rises with the
prevalence of the standard.* And that's just for the end users. These
"network effects" lead to "lock-in." It's what the Sun/Java story is
about—and it's another example of how the world is blurring.

Windows is the example Arthur uses to make his point. People select particular computers for the software that is available. As the number of software users increases, the value to the web of existing and future computer purchases rises. But software availability depends on developers, all of whom typically look for the largest markets. Thus, the more the market becomes saturated with Windows as an operating system, the more valuable it becomes to both programmers and software buyers. The battle between the Windows and Macintosh operating systems was decided long before Windows appeared, when IBM chose Microsoft DOS as its operating system. No one doubted that Macintosh had an interface that users preferred, but as PCs took off in corporate America, they were purchased by managers in the IBM web. IBM's market power established DOS as the standard, giving Microsoft the leeway to develop a more competitive product.

It was a self-fulfilling prophecy: Windows did become the standard. What's more, because the system involved a learning curve, users became locked into it for future upgrades and add-ons. They had become customers for life—though Sun's Scott McNealy might call them prisoners unto the grave.

Arthur credits the phenomenon of increasing returns to the heavy dose of Intangibles that is present in all software. We not only agree but go a step further. Whereas he claims only that information products are governed by this new law of economics, we think all offers will contain software and therefore will be governed by at least some aspects of increasing returns. Driven as they already are by two members of BLUR's trinity, Speed and Connectivity, the Intangibles that software provides can't be far behind.

The reason Microsoft enjoys increasing returns is not so much because its product is intangible but because the market receives a reinforcing benefit from having a standard operating system. The driver of the decision is not quality as much as compatibility. When does a market seek a standard? Whenever there is a need for Speed and where there is a use for high levels of Connectivity. In other words, everywhere, all the time.

Two simple examples: telephones and the English language in the business world. Surely no one thinks it's impossible to have a communication system other than the phone. Many people, in fact, prefer e-mail. Even short-wave radio has its fans. So why does everyone have a phone? Because we want to maximize our compatibility. It's crucial for us to communicate—connect—with as many others as possible.

The use of English in world business grows in the same way. Few believe it's the "best" language; although it may be the most adaptable, it has far too many idioms and exceptions. But starting with the defeat of the Spanish Armada in 1588, Britain's web of global commerce and the industrial power of the United States that followed tipped the balance for the business world to speak English. The clincher came with the combination of international air travel and the early dominance of the United States in software. The result: English is locked in as the standard. No one has the time to add another language for each new trading partner, especially as they try to expand globally into all manner of language zones. Once again, initial acceptance and the benefits of a standard together feed increasing returns, in this case for English.

Compatibility has been a concern in phone systems and language for a long time. We contend that, as Speed and Connectivity become even more central to our lives, the push for standards will happen in all kinds of other areas. It will also take on more urgency. How many phones, VCRs, thermostats, microwaves, and other appliances can you remember how to program? Whether your offer is tangible or intangible, the potential is there for lock-in—and increasing returns.

Not to belabor the point, the push for standards is part and parcel of the emergence of economic webs. Standard processes, platforms, components, and systems allow companies to connect *and* separate. Strong standards had to be in place before the most rudimentary steps could be taken in outsourcing, shared services, and remote teamwork. Broadening economic webs will demand new and extended standards.

Lock-in isn't all good, though. It can slow innovation, since the market is reluctant to abandon an existing standard. This is not unlike natural evolution, which builds and adds to an existing feature far more often than it invents new ones. American television viewers suffer poor color quality, for example, because way back in the early days of black-and-white TV, the United States settled on the NTSC standard. The rest of the world made later decisions and, as a result, enjoys better color.

Our belief in the need for a single standard clearly brings us up against the antitrust laws. We believe that this kind of monopoly creates the greatest value for the web as a whole. Remember, one standard doesn't have to mean one enterprise. This is the distinction between Windows and Java. But one company may not always mean monopoly in the sense of "bad for the customer"; so market share and industry concentration, the time-honored test of antitrust law, may not necessarily serve the public interest. We believe that the law will be rewritten to distinguish between the value of even a substandard standard (Sun's McNealy calls Windows "a hairball") and the abuse of the standard-setting position (the alleged poisoning of Java as a cross-platform standard or the alleged unnecessary bundling of Internet Explorer with Windows 98).

Real Markets Will Mimic Financial Markets

Diminishing returns giving way to increasing returns is just one economic oddity in the world of BLUR. Much of classic economics stands to fall similarly by the wayside. For example, the core assumption that the set of goods and services remains fixed for the period under analysis becomes laughable in BLUR. Concomitantly, some familiar economic abstractions will finally make the leap from theory to reality. The classical economics assumptions of "perfect information" in the marketplace, for instance, and the efficient workings of the "invisible hand" are actually in our reach, captured in the phrase "friction-free capitalism."[8] We're already seeing this in markets for computers, audio equipment, and cars. Like it or not, *the markets for what economists call "real goods and services" are more frequently behaving like financial markets.* Even in the recent past, Wall Street seemed to enjoy a special dispensation in how it dealt with pricing, risk management, product knowledge, feedback, regulation, value mind-set, and source of value. Not anymore.

Pricing

The first of these, pricing, is what most obviously sets financial markets apart. Call your broker and buy some shares of Ford Motor. Call back a day later to buy more. Can you imagine yourself saying: "What!? The price has changed?!" The fact is, we all know the prices of financial instruments shift from minute to minute, and we also know why: The players in the market are tightly connected, the instruments change hands quickly, and everyone gets instantaneous feedback of information. Imagine the same were true for the shampoo, the cough drops, and the khakis you buy. It's not inconceivable. Already, we think nothing of going to a gas station and finding the per-gallon cost different from the day before. Those updates used to be fairly infrequent; station owners could paint them on signs. Now, the signs are built to accommodate frequent alterations because the prices change weekly, if not more often than that. The same is happening in groceries. We hear that the coffee harvest is bad, so we expect the price to go up. Adjustments used to be year to year; now they ride in the slipstream of the news. In the world of BLUR, they may be continuous.

Real-time pricing is already happening in real goods markets for airline tickets, energy, long-distance calls, even electronics. For example, NECX is an online exchange for computer components that links more than 20,000 certified sources worldwide. Because it has real-time information on availability and pricing, NECX is able to offer dynamic pricing for more than 200 million electronic components. Customers get other benefits, as well, including side-by-side product comparisons, customized searching, a view of the stock in hand, and the ability to track their orders. How long can it be until Starbucks features a ticker quoting hourly changes in the price of Java Rich Roast? How long before potato chips are marketed like blue chips?

RISK

Real markets will come to resemble financial markets in several other ways. Notably, risk management has become an arcane science precisely because capital markets are subject to price volatility. As the prices of real goods become just as mercurial, we'll see the real-market equivalents of derivatives, insurance, diversification, and other risk-abatement tools. A simple harbinger of this is the guarantee offered by Tweeter, a chain of audio equipment stores in the northeastern United States. If you buy something from one of its stores, then see the same item advertised for less in a major local newspaper within 30 days, Tweeter will automatically mail you a check for the difference. Even more impressive, you don't have to do anything, such as send in a claim. When you make your purchase what you paid, along with your name and address, is posted automatically to a database as you leave the checkout counter. Over the next month,

Tweeter employees scan competitor ads for prices. In essence, there-fore, when you buy your headphones, you're also buying a price hedge—the value of the electronic offer has been enhanced by an intangible—financial insurance.

SYMMETRIC PRODUCT KNOWLEDGE

An important difference between real markets and financial markets has been in the levels of product knowledge that buyers and sellers bring to the transaction. In real goods markets, there's a notorious tra-dition of information asymmetry. As the buyer of a new car, you saw the sticker price, but knew it was a fiction. The problem was, you had no way of knowing the invoice cost paid by the dealer, which meant he held all the haggling cards. No such asymmetry exists in financial markets. The information exists for you to do a yield-to-maturity cal-culation as accurately as the pros can, and future corporate earnings are anyone's guess—the CEO's, the analyst's, yours, and ultimately, as moving shares prices will attest, that of the market as a whole. In fact, in financial markets, if you use asymmetric information, you could go to jail. It's called insider trading.

And where does product knowledge balance out in real markets? Everywhere. You don't even have to go to the library and dig up back issues of *Consumer Reports*; it's all on the Internet. If anything, a focused shopper is now better informed than the seller, who may not have studied competitors as thoroughly. Check out www.compare.net to scan product features and prices faster than you can dial the phone. Buying tires? Go to the TireRack Web site and you can learn more than you're likely to want to know. Have them shipped to the garage where the guy who *used* to be your best source of knowledge will install them. It's kind of funny: Now that the jig is up, note how forthcoming many sellers are becoming. Car dealers post invoice prices routinely. They even display competitors' vehicles, complete with stickers, alongside their own.[9]

REAL-TIME FEEDBACK

More accurate and available product information is fueled in part by faster feedback loops, another financial market characteristic appearing now in real markets. Feedback in financial markets, of course, is very close to real time. Quarterly earnings are posted, and share prices react immediately. Investors hear bad things about the quality of a holding and waste no time in dumping it. By contrast, feedback in real markets is glacial. Wondering whether the new model is playing well with customers? Stay tuned for the next sales report—due in three months' time. Hate the new menu at Denny's? Might as well write your congressman. However, widespread Connectivity is tightening those feedback loops. At Cisco Systems, customers dial up the Web site all hours of the day to access documentation and troubleshoot problems. Cisco tracks the usage continually and responds to people's biggest hassles with quick upgrades and advice. Supermarkets, too, are tightening feedback loops with frequent shopper programs. By analyzing the shopping baskets of their best customers, they can continuously tweak their merchandising mix for higher profitability. What's more, they can cater promotions to individual shoppers on a real-time basis. Buy Gerber's baby food, and the register spits out a coupon for Pampers.

REGULATION

Regulation, long a feature of real markets, is becoming as untenable there as in capital markets. The great landmark was the floating of exchange rates, an act that served to acknowledge that the markets were so connected and speculators' information so current and symmetric that currency could not be mispriced, not for all the gold in Fort Knox. This insight was mirrored in deregulation of the financial services industry; and now we've seen it come about in three real goods industries—telecommunications, gas, and airlines—all once regarded as industries immune to change. Disaggregation, lower barriers to entry, and increased competition ended that image, and players in each of these sectors have become ferocious competitors. The Speed of business, the Intangibles that drive it, and the Connectivity of the economy spell the end of the style of regulation that, like antitrust legislation, was built for the competitive models of the pre-BLUR world.

VALUE

The concept of value is one last area of convergence between the markets for financial instruments and real goods and services. Our mind-set toward value in real goods has always been oriented toward component costs. We traditionally looked at companies in terms of their balance sheets and valued them according to the stocks of goods and capital they owned. In financial dealings, by contrast, we assign value by focusing on the potential for future returns. *Future flow, not past stock accumulation, is the essence of financial activity.* More often now, this mind-set is migrating to how we value real goods. The best evidence is the growing disparity between the average company's book value and its market capitalization, a gap known as Tobin's Q[10] and described by George Gilder in the *Wall Street Journal* as the "index of the entrepreneurial dynamite in a capital stock."[11]

Consider DreamWorks SKG, the aforementioned movie production company spawned by the partnership of Steven Spielberg, Jeffrey Katzenberg, and David Geffen. As we described in Chapter 1, when the initial public offering was floated, the outfit didn't even own a copy machine, much less real estate, inventory, or anything creative that investors could pay to see. They didn't care. They charged in like starving velociraptors and established the company's value at $2 billion. Or consider the valuation of WorldCom at the time of its attempt to acquire MCI. At a time when AT&T and the Baby Bells had an installed base of network equipment worth $264 billion and a market cap about the same (for a Q of 1), WorldCom had about $5 billion in equipment but a market cap of $33 billion (for a Q of 6.5). That's a lot of Intangible value based on future expectations. It's also an incredible acquisition advantage, a dollar of tangible asset for just 15 cents and high expectations. But don't scoff; remember the absurdity of assuming a fixed set of goods and services? Your valuation sense should be tuned to the uncertain future, not the unreliable past. Too much risk, you say? We never promised you a BLUR of roses.

This new mind-set will soon encourage players in the real goods market to think about the source of value quite differently. Now the value we derive from the goods and services we buy is in the use of them; the source of value with financial instruments is in the trade. Art and antiques have, of course, always behaved like financial instruments, but today, secondary markets have opened shop on the Internet for used laptops, computer parts, and all kinds of accessories. We know of a case in which someone bought an Acura NSX—an $85,000 car—not for use but to hold as a speculation.

The table below highlights some of the differences between real and financial marketplaces.

TYPES OF MARKETS

CHARACTERISTICS		Real Markets	Financial Markets
	Price	Fixed	Floating
	Knowledge of the Offer	Asymmetric	Symmetric
	Feedback Time	Lagged	Real Time
	Value Mind-set	Stock	Flow
	Source of Value	Use	Trade
	Regulation	Possible	Unsustainable
	Risk	Eliminate through Design	Adapt and Hedge

Ask yourself where in your business the "real" characteristics are in force, and consider what would happen if they were replaced by their "financial" counterparts. What if your customers knew instantly when your component supplier dropped prices? What if the stock market saw your plant as a barrier to exit, not a barrier to your competitor's entry? What if your customers speculated in your product, changing price and demand when they changed their minds?

New Strategies for a Blurred World

We've noted that sound corporate strategy is founded in an understanding of economics. Given the emergence of economic webs, and the wholesale dismantling of traditional economics taking place right now, what's a company to do? It's no wonder that, in 1991, *Forbes* reported that corporations left and right were laying off their staff economists: "Like kings of old dispensing with their astrologers, big business is sacking its economic soothsayers," the magazine said. "Their stargazing proved entertaining and interesting—but not very useful."[12]

Here are a few thoughts on what the new economics is urging you to do. First, the importance of Speed means a shift from relying on prediction, foresight, and planning to building in flexibility, courage, and faster reflexes. And with Intangibles as the driver of value, your strategy must constantly focus on ways to increase the nonphysical component of what you make and sell. High levels of Connectivity mean that strategy can no longer be a matter of "us against the world." In the future, it will consist of early recognition of the right players to link with. Most important, it should revolve around the quest to set the standard. Finally, the arrival of real-time, symmetric information calls for the transformation of how an offer is sold and traded.

What are the implications of trying do all of this simultaneously, while the sands shift rapidly under your cornerstone? Essentially, BLUR demands—and enables—you to:

- Let the market manage your offer.
- Let the market price your offer.
- Let the market market your offer.

If this sounds more like an antistrategy than a strategy, it probably is. But let's look at each of these seemingly outrageous notions in turn.

LET THE MARKET MANAGE YOUR OFFER

Think about the role of the traditional product manager. She looks for applications for a new technology or capability, refines that capability to fit market needs, formulates specific offerings with clear value propositions, and determines which channels and channel partners are required for maximum effectiveness. Now ask yourself: Why is that person necessary? In a densely interconnected economic web, these things can happen spontaneously, efficiently, and far more effectively without a manager in charge.

Think again about Sun and its management of Java. By putting funds into a venture capital fund, it's priming the pump for applications to be found—not banking on discovering the killer app on its own. It's also letting the market, rather than some politicized capital budgeting process, allocate the scarce resources and developer talent to the most promising opportunities. Similarly, Netscape is making refinement of its software a role for the market to play. Its "bugs for bounty" program is a stroke of genius. Rather than paying its own developers to sweat over every line of code again, it incents users to find the problems. There's even more genius in the offer: For customers, the bounty takes away much of the sting of being bitten by the bug in the first place.

How else do providers involve the market in their offer? For years, radio stations recognized that rush-hour drivers, stuck in gridlock, were critiquing the traffic reports, so they began to ask their listeners to pick up their cell phones and contribute. News programs ask citizens for the same kind of help in alerting them to breaking stories. MTV goes one better, verging on "let the market manufacture your offer." Practically speaking, MTV is little more than a creative (and profitable) go-between for bands that want a showcase and fans that will do anything to connect with rock'n'roll. With rock videos (of bands), the MTV Beach Party (fans dancing to bands) and "Real Life" (fans when they're not dancing to bands), MTV is basically a channel for, by, and of the audience.

Cartoonist Scott Adams, meanwhile, is always ready to have readers of Dilbert supply him with examples of corporate mindlessness. In his case, the market is doing so much for him that he has been able to quit as an IS manager at PacBell, a job that hitherto had been his best source of material.

These examples show that the product manager is becoming a Webmaster. Chances are your market can do more for your offer—and even more so, your entire economic web—than you've been asking it to. Don't try to beat the BLUR. Join it.

LET THE MARKET PRICE YOUR OFFER

Pricing is a task you really have to let go of. It just isn't your job anymore. This is the lesson that car dealerships have had to learn as they sign on to the Auto-by-Tel network. As described previously, this Internet-based service allows buyers to locate the best deal in their area on the vehicle they want to buy. Literally every order is put out to bid to the network of dealers to which Auto-by-Tel connects in any geographic area. The price is determined by auction, by the market, every time. And it changes constantly.

Tweeter, also cited earlier, is another example. It revises its prices whenever a competitor is charging less. We've also talked about American Airline's real-time seat auctions. The point is: Accept that the market for your offer wants to behave like a financial market, too, and that the invisible hand is already dipping into your wallet.

LET THE MARKET MARKET YOUR OFFER

Perhaps your biggest leap of faith will be to recognize that the market can market your product effectively for you. It's probably already true that *only* the market can do so. This has been a long time coming, and has everything to do with the information overload each of us feels compounding every day as a result of media proliferation, cheap communications, and readily available marketing resources. (The current dark side of the BLUR.) As consumers, we feel as though marketing messages are assailing us like buckshot. As marketers, however, we know we can no longer depend on the single blast of a cannon to deliver our message. But increasingly overloaded customers are getting more irritable all the time. They have gone from manually zapping TV ads and hanging up on market researchers to installing TV sets with autozap and equipping their phones with caller ID filters. Their defense will only get stronger. What's a poor market manager to do? At Virgin Atlantic, they're waging war on the British Air–American Airlines alliance in a new medium: by painting "No Way BA/AA" on the outside of 747s. If only birds could read—and vote.

Disney is an exemplary case of what works now. Even while its next movie is still on the drawing board, the company is already striking merchandising deals left and right. It grants those rights to firms that will create marketing to fuel Disney's sales, as well. And, based on past success, it has the ability to ensure that the quality of that marketing will be on a par with its own. George Lucas goes further. His Skywalker Productions doesn't manage the market; it guarantees the standards that let the market thrive (just like Sun with Java). Every action figure, paperback, and T-shirt bearing Star Wars content is first assessed for consistency and then incorporated in Skywalker's record of the Star Wars universe. How is this different from Disney? Disney develops all the content itself, whereas Lucas lets the market contribute.

Again, strategy is all about determining your own company's relationship to its environment. In a world of economic webs, those relationships are profoundly altered. We've said that you can't tell where you end and the outside world begins, so let the market decide what it can for you—it will do a better job. It's not such a strange idea when you consider that the market has, all along, been managing your most compelling performance index: your market capitalization. With the blurring of real and financial, what else would you expect? Without a doubt, it's going to require some rethinking at a very high level. So BLUR isn't the end of strategy; it's just the end of strategy as we know it.

The Economic Web

Do you know your web? Who are the players? What do they want? What are their relationships? Who's generating how much value? Which players outside the present web could change the relationships? Will your place in the web grow or shrink?

Increasing Returns

Which parts of your business can offer increasing returns? Are you managing them differently? What is the role of scale in your business? Where does lock-in play a role? Are you adopting policies controlled by the market or the provider?

Financial Trading

Are you prepared to run your business by the rules of financial markets? Are you ready for your prices to float? For customers to have equal information to what you have? For value to come from flow not stock, and from trade more than from use?

CHAPTER 5

THE
ORGANIZATION
WEB

Run Your Organization by the Rules of the Market

The boundaries of organizations become so permeable that their identity blurs and recedes, while the importance of both smaller (individuals) and larger (alliances and Silicon Valley-like regions) economic units increases. Real-time organizations will cease being an oxymoron.

Since 1960, Let's Go has published a series of travel guides for the budget-minded. By 1998, there will be 26 regions covered, ranging from Sweden to New Zealand. But the company has only one permanent employee—the office manager. The 230 other staff members are Harvard students, who change every year.

In the days when Bill McGowan was building MCI into a respected long-distance phone company, he put the outfit through a major reorganization every six months or so, severing existing relationships so that his enterprise was encouraged to innovate.

At Morgan Stanley and various other investment banks, a team forms around each new opportunity, sees it to the finish, and then disperses.

Capital One Financial puts 14,000 new offers into the credit card marketplace every year. Most of them fail; nevertheless, the company is one of the most profitable in its field. A major reason for this is that every offer has a product champion but moves through the organization without the development of new structure.

In 1989, the Western world watched awestruck as the Berlin wall fell, taking down not only the regimes that erected it but the entire concept of a centrally planned economy. A lot of triumphant "I told you so's" were heard. However, many of those who gloated over the collapse failed to notice that in one significant respect, Soviet behavior not only had survived but was alive and well in their own backyard. The point here is that, managerially, much of the West runs its corporations like the Soviets ran their economy.

Did that get your attention? Let's step back and take another look at what went wrong with the economy that the Soviets espoused and foisted on their dominions and see what it tells about today's organizations. At the simplest level, the *apparat* of planning was unequal to the task of allocating investment, materials, and labor to fulfill the desires of a diverse population. Goods were scarce and shoddy, or plentiful but unwanted. Capital equipment stood idle for want of parts. People "attended jobs," accomplished little and derived no pleasure from them. The Russian aphorism summed it up: "We pretend to work, and they pretend to pay us."

But we're the technocrats. Shouldn't we admire all the computing power that Soviet central planners applied to solving the economic equations? Don't linear programs run our refineries effectively? What has gone wrong? Maybe if we just tried it with "real" computers, instead of the ones they put on the MIR space station. . . .

Adaptability

The heart of the matter is adaptability. Sure, linear programs can optimize a refinery, because it's a stable system: We know what the inputs are, what the desired outputs are, and how the process in between works; better still, that process doesn't change (or if it does, we have plenty of time to rework our mathematical model while we build the physical one). That's how planned economies worked, too: A statistical "input/output table" was created so that planners could list the output mix they deemed desirable, and then schedule all the inputs and intermediate goods.

But how do you solve problems with literally millions of interdependent variables, each of which can change at any time? At the same time, how can an economy adapt to continual flux in everything that affects resources and desires? Answer: You break up the problem into little local pieces, and solve them independently as best you can.[1] These pieces are called markets, and each solution is called the market clearing price. Market economies have been solving such problems since the pre-Christian trade routes—which is why Java is letting the market develop its product, instead of designing it all centrally.

But guess what? The United States and other Western economies continue to provide a safe haven for the command-and-control approach that marked Soviet economic management, not at the macroeconomic level, but in their management style. We've already said that strategy in the age of BLUR means that an enterprise must adapt to its environment, the economic web. But an organization runs a mini-economy, with thousands of continually changing variables. It needs to be every bit as adaptable as the economy in which it participates. Maybe you think your organization—the place you work, help manage, make the big calls—is just that: adaptable. Maybe you'll protest that it isn't run with a Soviet-style fixed input-output table. But then, think: How often have you heard somebody ask, "Budget final yet?" Actions aren't determined by a centralized decision-making body like the Politburo? ("Anybody got the table of approval authorities?") Organizations don't have people in unrewarding, unproductive jobs? (Oh, please.)

Until now, corporate organizations almost invariably followed the rules of power and status generally associated with bureaucracies, if not dictatorships. A market economy, on the other hand, runs by the rules of economics, which means relative freedom from arbitrary rules and hierarchy and the opportunity to react instantly to new information. This is why so many senior executives are constantly frustrated with their organizations: They are eternally out of step with an adapting economy because they are hamstrung by rules that act as barriers to internal adaptation. We're all familiar with the complaints and excuses—the boss won't like it; we can't get the budget for it; the person with the great idea doesn't have the clout to make it work. . . . Maybe we've even spoken such words ourselves. Real leaders, however, *want* greater adaptability than their organizations know how to deliver. Is BLUR the problem or the solution?

For the Soviets it was the problem. The drivers of BLUR precipitated the collapse. Speed, a property of the economic web described in the previous chapter, left those notorious five-year plans in its dust. Connectivity, whereby change in one corner of the market instantly affects every other corner, caused critical imbalances between tractors and turnips. And Intangibles, measured hardly at all by Soviet economists, put the planned economic output even farther from consumers' real desires. All the weaknesses of a planned economy are magnified a thousandfold by BLUR.

If you're not adaptable, BLUR will get you, too. Organizations that fail to replace bureaucracy and regulation with internal markets and other adaptive features—companies that resist BLUR, in other words—will fall like the wall and the Soviet economy. Discredited, laughed at, in rubble.

But for the company that sees the BLUR, it's part of the solution. Bill McGowan understood this. That is why he arbitrarily reorganized MCI every six months—not to solve a specific organization problem, but to quash political structures before they became intractable. This strategy opened the opportunity to form new networks. No single reorganization was intended to create the perfect new organization design (planning), but rather to give individuals a chance to self-organize and evolve in reaction to the changes taking place in the business around them (adaptation).

This chapter is about building an adaptive organization. First, this means that the organization web must run by the principles of the economic web. Economic activity obeys the same rules at the level of the economy, the company, and even the individual—BLUR is *fractal*. Second, ideas about adaptive systems currently being developed in the world of science can be applied to organizations. Take variety, for example: It's worth paying a price in efficiency for the *diversity* of thought that breeds innovation. Also, making *boundaries permeable* makes it possible for new ideas to emerge and an organization is most robust if it's unstable, at "the edge of chaos." Third, being simultaneously big and small, and creating a healthy churn within the organization are two strategies for building adaptivity.

LOOKING FOR THE FUTURE ORGANIZATION

The way desires are fulfilled changes radically when an economic transformation occurs, as it did when the industrial hierarchy replaced the small proprietorship at the end of the Industrial Revolution. Only as General Motors, Dupont, Standard Oil, and the railroads were becoming huge national enterprises was a new organizational approach required. We can expect that this pattern will repeat—organization change will follow the change in business. As enterprises fully embrace the economic web, a new organization model will also emerge, but since this will happen later, it's more difficult to know today what it will look like. But here's a clue: Learn how to build an organization that adapts to the economy as fast as the economy changes; in other words, build a blurred organization. As you do this, remember one crucial fact: It is not different from the economy; it's a part of the economy, subject to the same forces. Thus, your organization needs to run by the same rules.

Chaos theorists talk about the "fractal" property of many natural things, meaning they have the same structure at different levels. A tree, a branch, and a twig—the branching structure—are the same at all scales. That's because they are all parts of our connected system, performing the same role.[2] Likewise, the individual, the organization, and the economy, are all trying to generate value within a consistent set of economic rules.

The big idea, then, is this: The methods for value creation must be the same at the macro- and microeconomic levels—the economy and the enterprise. The structure of the economic web described in Chapter 4 should be the same as the structure of an adaptable organization web. This enables your company to be both a part of the economic web (a branch of the tree) and an economic web itself (a branch full of twigs). We'll see in the next chapter that this fractal property—using the same principles at all levels—applies to the individual as well.

The primary implication of all we've written about the economic web is that it applies internally, too. The CFO's "product" isn't a set of financial reports, it's an offer of information, analysis, and action; it's the same for the product engineer's design. The business units don't buy benefits management from the shared services organization, they exchange buying power, feelings of security, and individual histories for insurance, retirement benefits, and the like.

Your organization cannot be a set of little factions in a planned "economy." It must be run by the rules of the economic web, managed by the market, displaying increasing returns, and forming evanescent alliances. Since you can't define the boundaries of your connected organization from the outside, you shouldn't define them from the inside either. That's why the distinction between "intranet" and "extranet" is a step backward. It's like putting a blockage in every capillary, it's unconnected and therefore unhealthy.

Adaptive Organizations Are More Organism Than Machine

Before constructing the organization web on the principle of the economic web, let's explore some of the differences in mind-set between a designed organization and an adaptive one. In the preBLUR industrial view, the economy is a machine. Parts get designed, built, and assembled into a smoothly functioning whole that, once built, runs for its lifetime, mass-producing valuable output. In a machine organization, we fine-tune, restructure, and resize the place as if it were a mechanical device. Once we get the design right, we freeze it, and we intend to leave it alone for as long as possible.

The willingness to find comfort in such ruts is inherent in the logic of the business machine itself, not just organizational structure. The experience curve, for example, a by-product of World War II that emerged from data gathered about building airplanes, assumes that the productive task *will remain the same* while experience accumulates. If the task doesn't remain the same, you have to launch a new experience curve. This type of thinking means that all the strategies aimed at gaining market share are based on a kind of zero-sum game:

1.   The market and product boundaries stay stable.

2. The industry comprises a set of companies (or machines) to fulfill a demand that doesn't change its nature.

3. The overall approach to product doesn't change.

One who questions the applicability of this theory is Michael Rothschild,[3] the first to articulate in detail the economy in terms of biology. The irony is that Rothschild was trained at Boston Consulting Group, which made the experience curve the basis of strategic thinking for two generations of strategists. He pointed out that if the economy's objective is growth, there's an important flaw in the static model: Has anyone ever seen a machine grow? He put his finger right on it.

In the BLUR world, desires are evolving at a rate that doesn't allow the plan-produce-profit cycle to play out—or pay off. By the time the product becomes profitable, it has been supplanted by a new one. Such is the Speed of change. Sure, the old model still works, but the connected environment around it has changed, rendering it obsolete. Hence, Intel has three product teams working in parallel to ensure that there will be a microprocessor equal to the task of booting the ballooning code of Windows 9x. The experience curve drives the organization to efficiency only in a stable oligopoly. But in the BLUR world, efficiency doesn't drive the economy of the enterprise.

If you build your organization as a machine to do just one thing, everything is invested in the finely tuned process, and anything else is pared away. If you find out it's addressing the wrong question, it's tough to regroup, even if you are the chief executive. We can think of only one case where the boss of bosses publicly changed his mind—and the course of his company. This, of course, was Bill Gates' famous turnaround of Microsoft when he redirected it from a company totally focused on dominating the desktop to one that accepted that the Net was the technological next big wave. In Gates' so-called Pearl Harbor speech of December 7, 1995, he posed the new question, answered it, and directed the enterprise to pursue the new goal. Seldom is the change so definitive, the production method so intangible, or the leadership so incisive and decisive. What about the rest of us?

BLUR requires that we construct organizational structures that are designed for adaptability, not efficiency. That is why the ideal of "lean manufacturing," popular five years ago, is giving way to "agile manufacturing." We are transitioning to models of connected, growing systems from which we can learn.

For a decade or more, businesspeople searching for ways to adapt have been studying biology. Even though the suggestion that human-kind can learn something by studying an anthill often produces lots of chuckles, the biosphere has been successfully adapting to change for 4 billion years or so. Perhaps there's something we can learn and implement from nature's strategic moves. Kevin Kelly suggests, "When in doubt, follow the natural."[4] That means, think about your enterprise as an organism, something that senses the environment, adapts to it, and sometimes changes it.[5] Similarly, think of your business environment as a selection mechanism, dispassionately determining which

organisms fit. Fitter organisms come from breeding, combining existing ideas and new ones.

Now, let's see where this takes us. Be warned: We'll be throwing
in a little science to help illustrate these principles, but the focus of
each section shows how you can apply nature's adaptive systems to
managing and organizing your company.

Pursue Variety

Today, almost all of the corn grown in the United States is managed
by the Department of Agriculture. There are only a few dozen varieties, and they are bred and chosen for their properties of yield, disease resistance, and requirements of sun, soil, and water. What if the
environment were to change? A new disease, a shift in climate, or
more ultraviolet radiation could wreak havoc within this severely
limited gene pool. So the Department stockpiles hundreds of other
strains of corn, just in case of agricultural or climatic BLUR.

The John Deere company, whose products have an affinity to the
vagaries of nature, has recognized the value of this approach in its
organization. Deere was having trouble with its production scheduling
because of the rich variety of machines it turned out—1.5 million
potential models in all. The factory used an accepted, highly sophisticated approach to scheduling; but these plans produced a level of
overall quality that was unacceptable (remember those Soviet planners?). So Deere turned to Bill Fulkerson, an engineer with an eccen-
tric obsession: something called adaptive systems theory.[6]

Fulkerson went outside his organization to other adaptive systems enthusiasts, using the Internet for communication. Then he devised a solution to Deere's production worries based on one of nature's key adaptive approaches—sexual recombination—using software that emulates the workings of genes. Now Deere's computer breeds 40,000 schedules every night and selects the best of them by using this genetic algorithm. As in nature, the best of the 40,000 is not likely to be the best possible, any more than any single human being selected by nature is the best possible. The schedule is not optimal, but it is better than any that known optimization approaches can create. In biological terms, the schedule is like a species, adapting to the factory environment.

If Deere hadn't been tolerant of a wide "gene pool" in its engineering staff, or willing to take an open approach to the outside world (discussed in the next section), Fulkerson's innovative solution might never have been discovered. The lesson of this experience is that one of the most constructive ways to create adaptability within an organization is to increase the variety within it, and then ensure that the various knowledge and approaches mix with one another to generate new ideas.

The Santa Fe Institute (SFI) is another example of a group that chose intentionally to encourage "inefficient variety" to ensure innovation. Founded explicitly to explore what might be discovered by bringing together a multidisciplinary group of investigators, SFI has only three long-term faculty members. (Quite a trio, though: CalTech physics Nobelist Murray Gell-Mann, Stanford's increasing returns economist Brian Arthur, and MacArthur Fellow theoretical biologist Stuart Kauffman, from the University of Pennsylvania.) Everyone else is in residence or on a short-term appointment. The Institute conducts its work so as to bring people of cross-disciplines together daily. Afternoon tea, for example, is a part of everyone's routine, and any topic is fair game. Those who have been there have experienced a velocity of learning and an alteration of outlook they say is impossible in the standard academic organization, where the disciplinary boundaries remain unblurred—not unlike the functional boundaries in business organizations.

The speed of change on so many fronts—in science, in manufacturing, in the nature of demand, in the importance of other cultures—is now so great that closed, uniform organizations can't even hope to catch up. A couple of dozen varieties of ideas aren't enough diversity. To make organizations more adaptable, less likely to be blindsided, and more capable of creativity, bring in new types of people, or form teams that don't seem to be the most efficient—and won't be.

Is your company increasing its variety, and embracing it?

Have Permeable Boundaries

Have you ever tried to receive e-mail with an attachment from a client? Most corporations protect their communications systems with a "firewall," a security system designed to keep spies out, secrets in. As often happens with great intentions, however, these devices can produce hellish results. Their major failing is that they can't discriminate. They are just as likely to keep useful information out and impede those inside from reinforcing useful relationships. Remember that the Deere scheduling problem was solved by its engineer Bill Fulkerson, with the help of his friends on the Internet, an example of the benefits of crossing the organizational boundary. Would your company do that? Take this quick test—and be honest: What's your attitude toward sharing information with suppliers? Customers? Competitors? Not such a great idea, right? And, how do you feel about people who quit? Do you create an alumni club or shun them forever?

Most organizations have habits and structures that keep them at arm's length from the rest of the world. That's the way the machine economy is constructed, legally and philosophically. But though each organization is an economic web of its own, it is also an integral and intimately connected part of the larger global web. Where does the company start and stop? What does the boundary mean?

The adaptive organization benefits when the membrane surrounding it is permeable, just as the living cell is. This porous cell boundary is tuned to let in the oxygen, sugars, and other substances that the cell needs to carry out its specific assignment while excluding others. A muscle cell occasionally receives adrenaline, for example. Similar signals show up in the business world in the form of shifts in demand or innovations in technology. The boundary of both a cell and an organization must be porous enough to let in the information it needs. Otherwise, the cell (person) will be unable to act as the greater entity requires, and the body will suffer.

Investment banks, which are known for rapid change, provide a good example of the permeable boundaries necessary to grow an organization web. In order to respond to new opportunities in the market, these institutions must frequently modify their teams, adding people or affiliations with outside experts. Simultaneously, however, the banks must be able to modify their internal relationships. Porous boundaries—to the outside, and within—give them needed flexibility. Further, most individuals belong to multiple internal groups, making it easier to move people around.[7]

Why should we be willing to share details of how we go about our business with other companies? The answer can be found in Silicon Valley, where individuals readily share their views. They appreciate that spreading knowledge quickly increases everyone's learning, and they know there's more than enough opportunity to go around. People recognize that by giving away interesting information, they will likely attract other interesting people who will contribute more ideas than they take away. Contrast that with some traditional corporate cultures in which, internally, the folks in the marketing department won't give their counterparts in the finance department their data because they want to put the best spin on it first.

Encourage your people to talk, not only to each other, which should be obvious, but to people outside the organization. Reward them for establishing new relationships. Send them to meet groups of people with which the organization has no prior contact. Then the organization will start to adapt by itself, without your knowing it. Permeable organizations form external relationships easily and use them to bring in knowledge, talent, and opportunity. They lower barriers to circulation of value, and they maximize connection.

> *Are your colleagues and/or employees encouraged to exchange information with counterparts outside your organization, to cultivate knowledgeable networks, to seek solutions from disciplines not represented in your company?*

Instability: The Edge of Chaos

Admittedly, there's a frightening element in these recommendations. Do we want the organization drifting, changing, spreading and blurring, into other organizations, into the economic web? Won't that just preoccupy everyone with organizational change, and lead to a lot of chaos? Well, yes and no. Adaptive systems exist in a state known as the "edge of chaos,"[8] where they are stable enough to persist, yet flexible enough to quickly discover new solutions when a new problem arises.

Stuart Kauffman, author of *At Home in the Universe,* and a genius at designing abstract simulations that exhibit human properties, has created models to test these ideas. Kauffman's key variables were: the difficulty of the decision, the degree of interrelationship among the decisions, and the size of decision-making groups. Kauffman found

that decision-making effectiveness—the Speed with which the simulated organization finds a good enough solution—is crucially dependent on having enough chaos in the process.

Imagine a neighborhood in which everyone agrees that no adja-
cent houses should be the same color. How can such an "organization"
decide what color to paint each house? Kauffman first set up the organization as a single decision maker; that is, all decision makers had to
agree on the choices. The result? Frozen bureaucracy; no decisions.
Next he let every individual make his own choice. Since the decisions
were interrelated, however, as soon as one homeowner made a choice,
the neighbors had to change their minds. And when they did, that
tipped other changes. At this limit (Kauffman called it the "leftist
Italian limit"), no one will compromise, and the organization never
settles down—permanent chaos. The best results, found Kauffman,
occur when there are *almost* enough independent decision makers to
trigger permanent churn—"rather near the transition to chaos."[9]

So how many independent decision makers should an organization have? Is breaking your organization into business units, a popular
organization strategy since the 1960s, sufficient? Well, maybe for the
1960s. But BLUR, with its accelerated change and increased connectedness, pushes farther. In Kauffman's experiments, breaking up thousands of decision makers into groups of about six achieved the most
adaptive result. In a traditional hierarchy, this corresponds to an
organization run by first-level supervision. The management lesson is:
First, let the boundary of your organization be permeable, and move a

lot of variety into it. Next, let relatively small segments of the organization make their own local adaptations. Then the whole will take care of itself better than any central authority can. Remember those planned economies. It's when you start interfering with the chaos of the market that everything goes awry. The adaptive organization is constantly at the edge of chaos, open-minded enough to find new solutions, but not so fickle that it tears itself apart.

To summarize, your organization needs variety to be adaptable. You can increase that variety internally by bringing in new people with possibly conflicting ideas, and externally by ensuring that the boundary of your organization allows ideas in. Once you have access to that variety, you have to exploit it by instituting a rich decision-making process that opens it to use by many small groups independently, and lets each decision maker affect the others rather than trying too hard to achieve a consensus. This is an organization that will decide fast, exploit Connectivity, and use the Intangible value of its people—in short, a blurred organization.

Be Big and Small

Another characteristic of a blurred organization is that it becomes big and small simultaneously. Traditional wisdom has it that big companies are not adaptive; they are slow, bureaucratic, and unable to change. Small companies are thought to be quick, agile, and fast-changing. But in the economic web, small companies, too, have a problem: Connections take time, so if your resources are limited, you spend all your time nurturing the web and not enough adding value to your enterprise. In a BLUR world, you have to be both sizes. Organizations must be big to be capable of large-scale investments, with enough scope to encompass all that variety we talked about, while paying enough collective attention to changing global markets. And they need to be small: nimble, unified around a purpose, capable of paying attention to the details of each important relationship. How do you win this doubleheader? If it sounds impossible, remember, business schools preached that the concepts of mass production and custom-made were opposites. Then BLUR brought about mass customization. Are we seeing the birth of another oxymoron: agile bureaucracy?

How can you square this circle? The answer is infrastructure. Software, networks, processes, capital equipment—all can be considered part of the infrastructure. Remember, you don't have to own infrastructure to make it big. When CVS began opening drugstores all over the United States, local competitors started to close. Enter Bergen Brunswig, a distributor of drugs and health and beauty products. It saw the opportunity to provide its customers the same advantages of bigness that CVS gave its outlets: aggregated buying power, sophisticated software for order entry and pharmacy operations, and integrated logistics among them. Small store, big infrastructure. This gave the corner druggist the same economics as the big chains. Today, Bergen Brunswig is the country's number-two distributor, and the corner druggist is alive and well, often exacting a premium price for a different kind of service.

The implication is that if you become too attached to owning your assets, they become liabilities, because it's hard to leave them behind when someone else finds a better way to do what they do. In some ways, a start-up business has it made: It can rent space in an office park equipped with technical and administrative support, buy best practices from all over the world embedded in enterprise software packages, avail itself of market knowledge from specialized consultants . . . you get the idea. It can be big by connecting to the capabilities of others, and at the same time be nimble and adaptable because it's small.

But what if you're not a start-up; you're already big, and not sure how to get small. You have to determine where your assets can be a business in themselves; that is, figure out what should stay big and exploit it, not just in your own business, but in the economic webs of others, too.

The Morgan Stanley story cited at the beginning of this chapter is a good example. By creating the infrastructure that enables teams to come together and dissolve easily, its professionals can put their talents to the best use. The reputation of the firm attracts talent, the ongoing client relationships attract opportunities, and the financial capital provides clout in the market. But the work is done by small teams, using these and other infrastructure elements, such as telecommunications networks and databases, to create the most value.

Ernst & Young's consulting business runs the same way: By coupling all the firm's knowledge in its Center for Business Knowledge and implementing tools that help consultants help their clients, the firm enables each account team to adapt entrepreneurially to the needs of its client. The team is small, the infrastructure is big, the whole is adaptive. And that's not just a service delivery strategy; it's a human resource strategy. The best talent goes where the infrastructure provides the most support. So "big" infrastructure—analogous to the mass production capability of industrial companies—helps you get the crucial resource needed to be "small"—the people who can best use it, equivalent to the customization capability. Mass customization. Agile bureaucracy.

Soon, professional services companies will stop assigning staff to projects. Why? Because this is a planning mechanism for a process the market could carry out more effectively. Why not treat each client requirement as a job in the labor market? Let those who are interested bid on it. That way, the individuals involved, not the overworked staffing manager, can make the difficult trade-offs among cost, qualification, and development that are at the heart of managing professional services. Let's say your client insists on someone who has done the same job a hundred times before, but that manager wants to do something else for a change. The account manager is going to have to bid very high to get that individual to go through those motions again. Conversely, let's say you don't have the skills for a project, but wish you did; you bid close to zero and make yourself a bargain to gain the learning opportunity in the exchange. If you get hired, you get real training—on the job, not in the classroom—and experience that raises your market value next time. The account manager doesn't mind the low productivity of the neophyte because the client got the benefit of a highly motivated professional at very low cost.

The professional services companies will be among the first to bring the rules of the market inside. But, as suggested, some start-ups are already doing it. Teltech serves as a clearinghouse for technology expertise, putting corporations with questions in touch with the world's experts, usually university professors, on a given problem. Abuzz, a start-up, is developing Beehive, software that will let an organization put a price on the knowledge passed internally between individuals. People will finally be rewarded fairly for helping others in the organization who need their knowledge.

Stephen Jay Gould, the eminent Harvard paleontologist and pro-lific author, helps fill in the picture. He points out that, while we tend to picture humans as the ultimate success of evolution, the world belongs to bacteria and always has.[10] Bacteria comprise the greatest proportion of the biomass, the largest number of distinct species, and they inhabit every ecological niche from the sea floor to the mountaintop. There is more information in the DNA of the bac-teria inhabiting your body than in your own genome. Bacteria sur-vived not only the meteors that decimated the dinosaurs, but the planet's worst ecological disaster: the pollution by algae of the early, pure nitrogen and carbon dioxide atmosphere with the toxic gas oxy-gen, which killed many anaerobic species and paved the way for the mammals. In time, the bacteria adapted by using the new oxygen-breathing species that developed later as hosts.

How have bacteria become even more pervasive than Windows? Surely they don't plan, and they don't consider their competition, they don't attempt, as humans do, to change the environment around them. What they do is breed very quickly and mutate often. Both breeding and mutation are sources of new information; whether through recombination or random change, innovation is achieved. Then, of course, the environment gets "to vote" on the usefulness of the innovation: Most die; some thrive. Selection determines which few solutions survive to become part of the next round of innovation. Sounds like Austrian economist Joseph Schumpeter's definition of capitalism, namely: "Gales of creative destruction."

For an organization, rather than an economy, to take advantage of this bacterial approach, it will have to make some significant changes in its management approach. Bill McGowan's frequent reorganization of MCI translates into almost constant recombination. The trouble is, that's only half the story. Since the MCI reorganizations were almost exclusively internal, they didn't have the benefit of completely ruthless selection that weeds out all the ineffective relationships, structures, ideas, or people; and, maybe even more important, they didn't import very much by way of new information from the economic environment. That requires a constant churn *across* the organizational boundary and into the marketplace of offers, knowledge, and people. Consistent with BLUR, churning these Intangibles is the key to Speed of adaptation.

Churn Your Offer, Your Knowledge, and Your People

Capital One Financial extends credit to consumers for cars, houses, boats, and fun. The company began as the credit card subsidiary of Signet Bank, but grew so fast and profitably that it spun out to form a new corporation. In 1994, it went public at $1.07 billion, and by 1997 had a market capitalization of $3 billion.

Credit cards are hardly a new idea, but Capital One is making money by discovering tiny niches in the credit market more often and more quickly than anyone else—loans on cars older than two years in a falling interest rate market, for example, or second mortgages on beach houses in hot areas, and the like.[11] The right question is not how does Capital One do it, but how does it do it better than Citibank, Chase Manhattan, and just about everyone else in the industry?

The secret is churn. Capital One gives its people the power to create offers through recombinations and mutations, put them in the marketplace, and see what happens. In 1996, they made 14,000 offers, of which a mere 3,500 were deemed successes. But, in this intangible world of financial services, the cost of the 10,500 failures was relatively small compared with the big rollout of the successes, and the number of successes was higher than it would have been by designing, laboratory testing, and then offering ideas to the market for a true test. More important, each failure had intangible value: a learning experience in the market. In biological terms, each was a chance to prune some unhealthy information from the genome.

Capital One is churning offers through the marketplace to let it
select the best, and thus improve the breed.

According to George Overholser, Vice President of New Business
Development, one of the mantras at Capital One is "seed, select, and
amplify." This is how many so-called primitive species reproduce.
Rather than investing all the resources available in a small number of
live offspring like mammals, they plant "cheap" seeds like mushrooms,
or anything with spores. The wind, the birds, your Nike sneakers will
spread those spores to different environments. Most will die. But a few
will land on some newly changed corner of the environment and
flourish. Select these winners and amplify them by planting them all
over the place. Wait and see what happens. Replant the new breed
where it thrives. Repeat ad infinitum.

Capital One does not think in terms of maximizing its share of
market, but of maximizing its share of experience. If it can learn more
about the market more quickly than its competitors, it will make more
money from this business than they will. Every offer, win or lose, is a
vector of ideas between Capital One and its partners in the exchange.
To eliminate most of them in the laboratory instead of in the marketplace reduces the opportunity to learn. Keep in mind, the means by
which your organization learns is one aspect of your offer.

Offers can be seen as channels of communication within the economic web between your company's web and the larger web of the
marketplace. This is another aspect of "letting the market manage
your offer," discussed in Chapter 4. The implication is that if your line

of offerings is static, your business may not be adapting to changes in the marketplace. The offers at your company should be shorter-lived than ever before, and more of your offers should be new.

> *What's the half-life of your offers? How often have they churned in the past five years? Are you viewing the encounters between your offers and the market as a chance to gain knowledge?*

Part of churn is ensuring that new knowledge comes into your organization every day. The stock of knowledge must increase and mutate continually if the company is to adapt to the market. Two examples are Ford Motor and Asea Brown Boveri (ABB). Both companies accelerate churn in respect to best practices. At ABB, one of the duties of the extremely small headquarters group is to help disseminate the learning among the Swiss-Swedish conglomerate's 1,000 companies. The individual responsible for marketing, for example, selects and amplifies by touring the operating units looking for practices (seeds) that will be transferable, and by taking steps to move them to other divisions.

Ford relies more on self-organization. Each manufacturing manager has an annual "task," namely to improve productivity by 5 percent. The incentive, if we can call it that, is that the budget will be reduced 5 percent every year for the same output; it's the manager's job to find that 5 percent. To help, Ford has built an infrastructure to accelerate the transfer of innovations and best practices, in the form of an Internet-based database called the Best Practices Replication System. Each time an improvement is made, the plant responsible for that innovation is required to document it in the database. This information transfer may take the form of words, pictures, sound bites, or video, and generally includes not only the contact information for the manager who created it, but also for any vendor involved in developing tools, software, materials, and so forth. Wow! Look at all that Connectivity! Each plant at the receiving end is responsible for reviewing every item in the database and determining its applicability to its own operations.

At both Ford and ABB, the rate at which knowledge diffuses and is put into action is accelerated. This makes a big company act more like a small one because the time lag between when the first individual learns something and the last one with a need to know is informed is reduced.

Can you measurably increase the Speed of diffusion in your company? What knowledge is crucial to diffuse? Have you created the connections you need?

Neither offer churn nor knowledge churn is too radical a departure organizationally. They are a response to the need for Speed and variety, and they use the principle of permeability. Intangible and Connected, knowledge churns cheaply. Churning your people, on the other hand, is tough.

Students of what consultants like to call "large-scale change" all seem to come to the same conclusion: You can't make major changes in a large corporation without changing top management. Recent successes include Lou Gerstner at IBM and Michael Eisner at Disney. But this is like giving the mammals a shot at the top after the dinosaurs called it quits. Isn't there a better way?

Consider Branch Rickey, the general manager who built baseball dynasties in St. Louis, Brooklyn, and Pittsburgh in the '40s, '50s, and '60s. In addition to his many Runyonesque characteristics (he was fond of murky sayings such as "luck is the residue of design"), he can be credited with some pretty important management principles.[12] The most profound was to put a rookie in the starting lineup every year. This ensures that new skills enter the mix on an annual basis, that every veteran perceives himself as fighting for his job every year, and that the team can't get old all at once.

Is that ratio enough? Does your organization go through an annual change of 11 percent of the starting lineup? (We're not talking pinch hitters, coaches, or batboys here.) Would that 11 percent be enough? How much new blood do you need in your web? How vulnerable is your business to the Internet-equipped Generation X team that doesn't see all the barriers to change that you take for granted? What could you learn from them? If you want to avoid the years it takes to rebuild teams that don't continually churn their people, you'd better make sure that you're perpetually absorbing the new ways of thinking. As important, make sure you take them seriously—extinct is forever.

If people are the lifeblood of an organization, consider this: The average human red corpuscle lives 10 weeks. That equals a 10 percent churn every week in your own life blood.

How long does the average person stay in your organization?
Is this number shrinking? Can you use connections to the
outside world to create virtual churn and shrink your stable
core? Are you worried that people won't want to work for
an organization that believes in churn, whether internal (à la
MCI) or external? Remember, the Santa Fe Institute is all
churn, yet it is viewed as one of the top five laboratories to
work at by academics, who generally live in the almost
churn-free environment of tenure.

One of the key tools for creating an adaptive organization is
churn—of product, of knowledge, of people. Don't let any one of
them stagnate. Instead, promote obsolescence, measure half-lives,
induce recombination. Are you keeping all these things moving?

Are we sure about all this: Blurring organizational boundaries,
churning people, running the company by the rules of the market?
We are, and our convictions are strengthened by the phenomenon of
Silicon Valley. The cliché is that in the Valley you change jobs more
often than you change parking spaces. The reality is that talent is
mobile, companies do form around projects, job tenures are short,
and adaptability and economic value-added are high. This is not
about Netscape, or Marimba, or The Idea Lab as companies, but
about the Valley itself. The companies are able to get small by relying
on the shared infrastructure. Sun's Java venture fund probably would
not have happened if the company, along with Kleiner, Perkins,
Caufield & Byers, and a bunch of others had not been a part of a

highly connected economic entity, namely Silicon Valley itself, and its counterparts in the Flatiron District in New York City, Silicon Gulch in Austin, Texas, and others to come. This is the phenomenon Kevin Kelly called the "hive mind."[13] So where does the company end and the web begin? It's a blur.

To close, remember how we opened this chapter: The organization is part of the economy. The economy is fractal; that is, the same structure and principles operate at every scale. Chapter 4 described the economy-level structures that govern strategy in BLUR. This chapter applied the same idea at greater magnification, examining the implications for the firm. Once the market rules the organization, it will be apparent that the distinction between an organization and the rest of the economic web is a blur, and all those who work are but players in it. Now it's time to step up the magnification again, and see how the individual will blur. That's the subject of the next chapter.

> *What is decided by command in your organization? How could market forces do it better? Do you measure the lag between market changes and your response? Your churn of people and knowledge? Your variety? What barriers do you erect to preclude permeability? What makes you big? Small? What are the barriers to becoming an "agile bureaucracy"— for you, and for others without your assets? How could your assets be better used by others?*

Resources

Economists have long spoken of land, labor, and capital as the inputs to an economy. Although land is losing some of its relevance, otherwise it is not changing much. Labor and capital, on the other hand, are changing beyond recognition. Labor is no longer thought of as hours of undifferentiated wrench-turning, but as talent, not so much to be hired as to be applied to the issue of the moment. Hence our discussion of people focuses on the nature of the exchanges in which this talent will be the most valued resource. Capital, meanwhile, is shifting from an asset to be accumulated to an exposure that must be hedged, churned, or discarded.

PART IV

The

BLUR

of

Resources

CHAPTER 6

PEOPLE

Manage Your Stock Price and
Your Career Will Take Care of Itself

Connected individuals and their knowledge, not
the corporation, are becoming the key organizing
unit. The market is developing ways to capture,
measure, trade, and reward them accordingly.

The business of baseball—and all professional sports—changed forever the day in 1969 that Curtis Flood, an all-star center fielder, refused to report to work. Traded by St. Louis to Philadelphia, he argued in court against the reserve clause in his contract that gave the Cardinals the power to send him wherever it wanted. Flood didn't win his case (it made it to the Supreme Court), but neither did he end up in Philadelphia. A few years later, two other players took up the same fight and this time, the reserve clause was overturned through arbitration. Free agency, not baseball, has been the name of the game ever since.

Rock star David Bowie surprised the entertainment—and the investment—world when he announced he would float a personal bond issue. Ownership of the bonds entitles investors to a portion of Bowie's future income stream in two key areas: royalties on previously recorded material and receipts from future live concerts. The entire issue was snapped up within an hour, for over $50 million.[1]

During the early '80s, it was said to be cheaper to find oil on Wall Street than in the oil field. Today companies seem to feel that takeovers are the cheaper way to find employees. Cisco, for instance, uses acquisitions as a way to buy product teams because it takes too long to assemble them from the ground up.

What happens when Connectivity, Speed, and Intangibles converge in you? When your ability to connect means you can work for a company 2,000 miles away—or for five companies, or for a hundred? When your talent—your Intangible value—is so visible it can be priced on the open market, and even "securitized" like David Bowie's? On the flip side, and far more frightening, what happens to recent graduates five years from now, when 85 percent of their college-level knowledge base will be obsolete? Get ready for all this, because it's a whole new world. This is when BLUR starts to come home.

You already know that life in the BLUR is going to be very different for world economies. And it's going to present unprecedented challenges for businesses and organizations. It's going to make life a lot different for you, too. On a deeply individual level, you will have to reconcile some complex paradoxes, adopt some radically new perspectives, and change some near reflex-level behaviors.

At least five things are blurring and they're all significant:

1. Let's start with the distinction between you as independent entity and as member of society. For centuries, politicians and philosophers have debated the degree to which people will and should give up personal freedoms in exchange for the benefits of community. They've haggled over which end of the spectrum is more "natural." They've argued over which is the more important to defend. The ideological battle spread into business a long time ago and still rages every day—Microsoft as the libertarian ideal versus Microsoft as the Evil Empire—but look what's happened in the

real world. Paradoxically, people are more independent and atom-
ized than ever by virtue of their being so pervasively connected.
Connectivity enables you to seize more independence *while independence motivates you to become ever more* connected.

2. Another big BLUR involves your new relationship to the market-
    place. The most valuable thing you have is your knowledge,
    which includes all the relationships and other intangible assets
    you've accrued over the course of your career. That's valuable to
    other people, too, and you can sell it. In fact, you can then turn
    around and sell it again, or sell it in a thousand directions at
    once. After all those sales, you'll still own it, but it won't be as
    scarce or, perhaps, as valuable as it was. So, how do you behave
    at lunch with a colleague? Provide unique knowledge to make
    her relationship with you more valuable, so you can get more
    information later? Or keep your powder dry, and avoid diluting
    the value of your knowledge? The Connectivity imperative
    moves you toward disclosure to strengthen the relationship, the
    value of Intangibles suggests careful consideration of the half-life
    of your knowledge, just how unique it is, and the possibility that
    it is more valuable to someone else. *Either way, here's the proof
    that your accumulated knowledge can be profitably shared. But you
    must manage this valuable intangible asset carefully.*

3. Another important BLUR is the disappearing line between you as
    laborer and you as capitalist. For years now, most companies have
    been telling their employees they should act like entrepreneurs.
    Some companies actually mean this, and even help employees to

do so. A few even change their resource allocation ways to mimic the workings of a free market. Variable pay based on performance—once reserved for commissioned salespeople—is now in place throughout all kinds of organizations. But this variable pay has a twist, because it's often linked to how the overall business performs. Once again, independence blurs with interdependence. Meanwhile, the typical business is more frequently staffed by freelancers, consultants, contractors, and temps—people who rarely forget they've got their own business to run. If everyone is to become a "company of one," is the noncapitalist laborer now obsolete? In your case, yes. *Take up your company's challenge to be entrepreneurial, whether or not the boss makes it easy for you, but insist that you be a real capitalist, not just a psychologically oriented one.*

4. The first of our two remaining BLURs is already familiar but will more often mark your relationship with any organization you join: the BLUR between work life and home life. People used to demarcate these two areas of their lives rigidly, not just with timeclocks but even down to the tenor of their conversation. More often than not, an executive would "put on his game face" walking in the office door. Hour-long commutes home served as the decompression chamber. Now, of course, work goes home with you. Managers check their voice mail while the pasta is cooking and compose e-mail while the baby naps. Even Sybil couldn't switch personalities that fast. *For better or for worse, in the blurred world, the "work you" and the "home you" have to meld.*

5. Here's a final wrinkle to BLUR as it affects your work life. Not
    long ago, a Wall Street firm launched its most concerted and
    elaborate marketing campaign ever, targeting it at 50 men and
    women with whom it most wanted to shake hands. The firm sent
    customized mailings based on database records of the individual's
    personal preferences, and wined and dined them over a summer-
    long series of seductive events. Did all this land any lucrative
    M&A or IPO work? That wasn't the purpose. This was a recruit-
    ment drive. *The point is, you're not only a customer for a firm's offer
    on the resource side, you're also a customer on the desire side.*

Everybody's a Free Agent

It's a common saying that work just isn't what it used to be. Think
about the traditional "organization man," the kind parodied in "How
to Succeed in Business without Really Trying." That was a life in
which you joined a firm in your twenties with the main ambition of
rising through the ranks. You might jump ship at some point and join
a competitor, but that was drastic—if not downright ratlike, as the
metaphor implies. Much better to stay solidly and loyally onboard,
and accrue all the advantages of seniority. It wasn't unusual at all for
a man (of course, a man in those days) to do his life's work in the service of a single employer. Perhaps both your grandfathers did exactly
this.

Your father may have been more prone to job jump, but he was
still stable compared with today. His jobs probably lasted for more
than a decade each, meaning he might have worked for three compa-

nies over 40 years. Now think of how far you are into your working life, and how many times you've made a move. If you're average, you've changed employers every five years. Meanwhile, the numbers of years we actually work is growing along with the average life span. Many of us will keep working into our seventies. That means you may work for 10 employers. At what point do you realize you've blurred into being self-employed?

People don't "jump ship" anymore, either. They "make moves." This signifies a managed, strategic pattern. Far from a desperate act of last resort, the worker is taking an informed step to advance her career. Rather than leaving yourself at the mercy of a firm's particular hierarchical system, its pay grades and politics, you now have to manage yourself as a business. You're becoming a free agent.

ENTER THE FREE AGENT

As a term, free agency originated in the world of sports when Curtis Flood battled to have a say in his destiny. Like other players, he was, in effect, indentured to a particular team. If he moved, it would be somebody else's decision, and he'd have no more choice in which team he joined than in the uniform he wore. Note that sartorial overlap with the organization man!

Once free agency did come about, it turned the business of sports on its head. The power equation that used to characterize a player's relationship to his team reversed. Exceptional players are even dubbed "the franchise," removing any doubt about where the power really lies. At the mercy of their best players' demands, owners have agreed to ever rising salary demands, increases that also cause far greater pay disparities among teammates.

Curtis Flood wasn't the first. The same kind of revolution happened in the world of entertainment. In Hollywood, the big studios—MGM, Warner Bros., and 20th Century Fox—used to be vertically integrated autarchies. Actors and actresses had contractual affiliations to studios that left them very little personal control over the work they did. The power began to change hands when Charlie Chaplin, Douglas Fairbanks, and Mary Pickford recognized their value in the marketplace and formed United Artists. Now, the "franchise" stars rule, the studios are distribution companies, and the agencies that represent the talent are becoming the most powerful corporate entities in the industry.

Now Big Business is stepping up to the same plate. Its superstars are the celebrity CEOs, the buyout specialists, and the turnaround champs. The widely publicized salaries, perks, and total compensation packages given to these people are now lighting the eyes of middle managers, who, too, are beginning to demand lush signing bonuses. The same expectation has even trickled down to first-time job applicants. One résumé-screener we know reports shock at the tenor of the cover letters he gets from new college grads: "Eighty percent are all about 'my needs from an employer' as opposed to 'what I have to offer,'" he says.

So where does your managed self fit into the BLUR? The security you're looking for—at home, at work—most likely lies in your own skill base and how you wield it in the open marketplace. You may even perceive that there is greater security in part-time than in full-time employment, since you might be incited to build recognition in a broader market.

If you don't believe us, take the word of the world's most prominent management expert: Scott Adams, creator of the Dilbert cartoons. *The New York Times* reports that "Mr. Adams thinks that in the future, more people will be working as he is, part of a phenomenon he calls 'boss diversification.' Mr. Adams goes on to explain: 'The worst risk that you can face is to have one boss, somebody who can make your life miserable and then decide when it is time for you to go,' he said. 'It used to be that was the safest thing, but slowly people are realizing through downsizing that that might be the worst risk. The more customers and clients you have, the safer you are. People are going to gravitate to what is safest.'"[2]

YOU AS A FREE AGENT

If you've ever changed jobs to make more money (and we're willing to bet you have), you've acted like a free agent because you permitted the marketplace to value your worth. The question in the increasingly blurred world is: How frequently should you make such moves? In the sixties, if you held more than one job per decade, it sent the wrong signal. In the eighties, two. At the turn of the century, who knows? The point is, changing jobs today doesn't raise eyebrows. Job changes indicate marketability and value—not incompatibility—and it's up to you to manage them successfully.

Any good manager takes care of today and tomorrow, and that's what you owe yourself. One way to do this is to pretend you're a freelancer. (Or, if it sounds better to you, a consultant.) Think of the organization that's giving you a regular paycheck now as your biggest client of the moment. But there's no guarantee you'll have that client tomorrow. This means that every day must become part of an ongoing marketing process for your services. And because of the impermanence of work, be careful not to place all your eggs in that one basket. In other words, even if you have a great job, make sure you have a second one in the wings. Better yet, do some work to promote your marketability and value. Adaptability matters for the individual too.

The most challenging aspect of being a free agent is that it puts the onus on you to understand where the market for your skills is heading. And, if it's about to crater, it's your responsibility to gain new skills. Therein lies the rub, of course. For many people, it's much easier—more comfortable, less frightening—to be told which knowledge base they should bolster and what kind of personal development to undertake. Remember, making a move only because that's where the big money is could exemplify bad management. A better move might be to opt for less money in a situation where you'll also be picking up the next set of skills you'll be needing—a different exchange.

Of course, you could try for the best of both worlds. Instead of agonizing over whether to stay put and work with what you know, or quitting to take another job where you'll learn something new, why not expand your skills as part of your current job? This has proved a great tonic for many businesses. It pushes them into new areas and helps to attract new talent. At General Electric, for example, employees can apply for a temporary "Bridge Assignment," which allows them to move to a different department in order to learn new skills. The program's benefits are widespread: new challenges and learning opportunities for employees and the facilitation of cross-pollination of ideas throughout the GE organization. Much of the innovation and cross-fertilization in Silicon Valley is also due to these forces.

To repeat: Rather than being managed by the organization you join, you manage its contribution to your career. This is why free agents seem to step up their loyalty to their profession. Programmers in Silicon Valley identify much more strongly with what they do than with their employers, just as doctors and lawyers make their chief allegiance to their profession. In such cases, continuing education, periodic recertification, even disciplinary actions, are mandated by professional associations rather than individual employers—and so, too, is much of the networking you'll need to do.

Expect a few other things to change as the market increasingly regards you as a free agent. Number one: Intermediaries will gain power. Creative Artists, the Hollywood agency founded by Mike Ovitz, became a force as actors asserted their free agency and looked for expert representation to the market. In the same way, we'll see a whole new class of job brokers and talent scouts arise. Manpower on steroids. It's likely, too, that their payment scheme will shift, from a cut paid by the employer to a cut paid by the talent.

Are there any other lessons to take away from the pioneers of free agency in sports and entertainment? Here's one to contemplate: The stars today in both those industries are actually making more money from "ancillary" activities than from their core work. Endorsements, celebrity appearances, merchandise licensing, book sales—these are where the real money is, but access to them is limited to the best in the field. We're assuming you're the best in your field. So start thinking: What's the analogous set of activities? Where's the real money?

Securitizing Individuals

As value shifts to the worker in the connected economy, it's inevitable that financial speculators will want some of the action. Over the next decade it's safe to predict that Wall Street will devise new instruments to develop, measure, evaluate, and reward the knowledge and experience of individuals. He or she will be the investment vehicles of the twenty-first century just as small companies were in the twentieth and large ones in the nineteenth.

It's already happening. As already mentioned, entertainer David Bowie issued the equivalent of bonds securitizing his future earnings. And in London, a 24-year-old aspiring actress named Caroline Ilana made the news when she offered shares in herself. It was a novel way to raise tuition funds fast: She promised investors a piece of whatever she makes in the future in return for cash on the table today. The scheme succeeded beyond her hopes. Among those to buy shares were composer Andrew Lloyd Webber and actors Bob Hoskins and Emma Thompson. Play with this idea: Conceivably, her shares could find their way to a secondary market where strangers, tiring of her failure to get parts, could pressure her into other forms of work. And any success she enjoyed, meanwhile, would be a double-edged sword, because the higher her value rises, the more it will cost her to buy back her own shares. The notional lesson: Stay in control of the securitized you by holding on to at least 51 percent of your stock. That way, nobody can push you to do anything you don't want to do.

Of course, a secondary market for any kind of shares seemed just as problematic back in 1792, when the New York Stock Exchange opened for business. The concept of a "publicly held" company is far from intuitive; and even as the idea took hold, it seemed applicable only to very large businesses. Even today, the NYSE trades primarily in companies with assets of at least $40 million and no fewer than 2,000 shareholders. Decades passed before investors could buy and sell shares in smaller companies, first at the American and regional exchanges, and later at the NASDAQ (National Association of Securities Dealers Automated Quotations), which handles over-the-counter trades of what are usually small and thinly traded companies.

These days, the defining line between these exchanges is far more fluid. Microsoft, for example, trades on NASDAQ. The bigger point is that greater Connectivity—telephones, programmed trading, continuous real-time stock quotes, and PC software that lets you trade from home—has enabled the market to expand its presence in everybody's life. It supports barter among far-flung individuals, the purpose for which money was invented. That investors can now take a flyer on the intangible value of individuals is, in fact, just a small, logical next step in the blurred world.

Let's have some fun with this. Let's imagine some celebrity stocks we wish existed—provided that we could get in on the ground floor. Okay: Michael Jordan. That's obvious. How about Bill Gates? We're buying. One last one? Oh hell, why not—Elvis. It's not too difficult to imagine a time in each of these men's lives when they might have made a decision to go public. We'll follow through with the Jordan example.

As a sophomore at Laney High in Wilmington, North Carolina, Jordan might have needed spending money. But taking a part-time job after school would have meant less time shooting hoops. Rather than undermine his future as an athlete, the teenager might have sold enough shares in his future income stream to cover a new pair of sneaks, for example. Would he have had any takers? Perhaps among college scouts—or would that constitute insider trading? (Good thing the idea of "amateur status" is gone—that's beyond blurring!) Fast-forward a couple of years to the 1982 NCAA championships when Jordan, in his freshman year at the University of North Carolina, helped lead the Tar Heels to victory. It's already obvious this kid is NBA material. A good investment, in other words, even allowing for such risks as injury and early burnout. And Jordan, of course, could hedge his own bets, laying off some of the cash an offering would bring in by investing in other players—or industries! Quarterbacks were a hotter sector at that time.

Now let's say it's 1998. You're Michael Jordan and you're considering building a mansion, complete with basketball court and base-ball diamond. You could pull your cash out of the stock market, but that keeps soaring. The cheapest source of capital might be a bond issue, backed by your own assets and earning power. . . .

Like stocks, it's easier to think of the superstars you wish you could have invested in rather than all the "also rans" you're glad you missed. On the other hand, what would an infusion of capital mean to any one of the legions of computer programmers who join the workforce every year? Would the economic web be better aligned as a result? What might an early infusion of capital have allowed them to do? Nomura Asset Capital Corp. has created an entertainment-lending division in which it hopes to raise $1 billion to provide "unusual loans" to musicians, actors, and studio executives. Each of the loans is secured on the artist's future earning streams, and is then bundled by Nomura and transformed into asset-backed securities to be sold to investors.[3]

The point of all this is that, just like corporations, individuals will take advantage of securities markets to raise the capital they need to finance their goals. It has been our experience that when first presented with the concept of individual securitization, people initially envision the worst: naïve young people exploited by fast-talking moneymen, terms amounting to enslavement, a general dehumanization of workers like that of actors on studio contracts or the blues singers and rock stars discovered and sewn up by unscrupulous agents in the forties and fifties. But those fears reflect the power balance of the past. In a blurred economy, where knowledge is a greater source of value than land or capital equipment, where what you know is more important than what you own, and where the reputation of the people you're dealing with is an open Powerbook, most of the chips are in the hands of individuals.

THE BUYER BE WHERE?

A market in which to trade the securitized likes of Michael, Bill, or Elvis may not exist right now, but it surely will in the near future. The free-market reason: There's a demand, on the part of individual sellers and interested buyers. As a result, a whole set of mechanisms will arise to support exchanging securities in individuals.

One already exists as a computer simulation on the World Wide Web. Wall Street Sports (http://wallstreetsports.com) is a 24-hour commodities market where more than 17,000 people trade shares in more than 700 athletes in baseball, basketball, football, hockey, and golf. When you register—for free—you're given $1 million of funny money. Lose all of it and you're out of the game; but the top trader has amassed almost $7 million. What's a game one day is often a professional sport the next—whether it is played in Madison Square Garden or on Wall Street.[4]

As a real market develops, no doubt the first need to be met will be for auditing services, to attest to the accuracy of individual income reporting. That will give investors sufficient confidence to buy. At the same time, sellers will be looking for guidance on debt structuring and projected cash flows; as a result, new branches of investment banking will arise. Growing volume will necessitate standards (and therefore standards boards) and exchanges (no doubt Internet-based). Soon after will come sources of ratings and analysis—the equivalents of Moody's and Morningstar. Investors will want broad information, like the outlooks for different professions and the numbers of graduates in different fields, as well as detailed information on particular individuals.

Different investment philosophies will be used, with different ones holding sway at different times. There might be a Myers-Briggs school, for example, that bases buy recommendations on tests of psychological attributes correlated with success. A demographics school might look for patterns of performance in geographic regions, age groups, or socioeconomic classes. Some will always focus on current performance: Is this person getting consistently good performance reviews? Many investors will look for solid, reliable returns from steady performers. Many others will look for the bargains and thrills of betting on dark horses.

Given the eventual abundance, relative low cost, and high risk of securitized individuals, mutual funds are inevitable. Funds could focus on sectors—say, entertainers, or even stand-up comics—so that a few winners could offset a lot of losers. They could focus on age groups or other demographics. Imagine a Harvard Class of '99 Fund. Or they might focus very narrowly on a company. Investors might even be more interested in a Microsoft Employees Fund than in Microsoft stock itself. In fact, one might replace the other.

Regulation and legislation are equally inevitable. Age limitations will be set early on, and it will surely be true that people will not be able to sell shares in their dependents, or in-laws, tempting though that might be. Disclosure and communications standards will be established, as will governance structures. In this new world of securitized individuals, what is the equivalent of the board of directors? Surely it could be a valuable source of career advice and networking contacts. But what power might it have to veto lifestyle decisions made by the individual? Or to replace other key managers, like one's trainer or therapist?

Scratch the surface on this prediction, and it quickly becomes clear there will be broad and deep implications. Making individual incomes into public information will cause major disruption to long-standing social conventions. And it's possible that diffusion of risk and accountability to shareholders will affect an individual's work ethic—although it's not clear just how. Individuals already act differently when they "issue" debt, so who knows what the behavioral response will be when they issue equity securities? After all, Curtis Flood's case eventually returned the "equity" interest in a player's career to the athlete—who can now trade it. Owners have turned this into an opportunity to provide performance incentives. These individual behavior changes will accumulate to create broader societal evils and goods. In a world of securitized individuals, what, for example, happens to income distribution? As a minor point, what happens to tax policy? As a larger concern, how do the missions and mechanisms of institutions—like the Internal Revenue Service—change?

Understanding all the implications of individual securitization will take time, but it's clear already that some forms of development will come about. As an individual preparing for the near future, you should already be putting the pieces in place with an eye to your IPO. That means adopting a different perspective on the job you hold now, and on the work you will do next.

The Individual as a Free Agent in the Connected Economy

Free agency and, ultimately, the securitization of individuals are moving us in the same inevitable direction: toward the direct connection of the individual and the economy. In other words, we are seeing the role of the organization as a go-between diminish, as individuals have greater ability to participate directly in the larger sphere of economic activity. This is very similar to the pressures on intermediaries on the desires side of the economy. It used to be almost unavoidable in a consumer goods supply chain that a third-party distributor be in place, whose job was to break down truckloads of product from a manufacturer into smaller, multiple shipments to retailers. Thanks to greater Connectivity and the drive to reduce tangible assets by speeding up inventory dispersals, such physical distributorships are disappearing. A Colgate-Palmolive can ship directly to the retailers who are its customers. Disintermediation is also evident in personal finance beginning with the ATM, and now, online financial management from PCs. By allowing customers direct access to their accounts, banks and brokers have been able to all but dispense with the traditional intermediary—the teller or registered representative.

The organization has always been the intermediary between the worker and the market; it consolidates the labors of many contributors and allocates the product of that combined labor across buyers.

This will still be the role of organizations; they will continue to exist to coordinate input and aggregate demand. But for a given individual worker, the role of any specific organization will be far less central. It may still be a useful, efficient channel to the market, but it certainly won't be the only one.

NODE THYSELF

What we're talking about here is a reconceptualization of the individual as, essentially, a node in an economic web. A node is, of course, a point connected in a network to multiple other points. We used to think of companies in the economy as being these points, irreducible for practical purposes. Now, because of more pervasive Connectivity, economic systems operate at a more granular level, and we have the ability to see them on that level. Individuals become the nodes, connected to one another. Connectivity makes it possible; the need for adaptability makes it valuable.

One great benefit of this reconceptualization is that we can finally get past two false dichotomies that have shaped our thinking for at least half a century: first, the distinction between laborers and consumers; and second, the distinction between work and life.

The first of these artificial distinctions—between the laboring individual and the consuming individual—is a mainstay of classical economics. Chalkboard diagrams of how an economy works invariably show a flow of labor from households to companies and a separate flow of output back to households, making it pretty apparent that laborers and consumers are actually one and the same. Yet economics treats the labor market and the product/service markets as separate entities, rather than as two roles of the same individuals. In other words, each of us both works and eats, in a more or less sustainable balance. It's the way that businesspeople draw the value chain that confuses them: They insist on a linear model, with their inputs on the left, the company in the middle, and the market on the right. Reality is much messier than any linear chain. First, it would be more accurate to draw the direction of the value chain from right to left—you can't generate value with offers that don't fulfill desires (which is why it's the value chain, not the fortune chain). Second, each of us is an economic node in a large, complex, adaptive system, namely, the economic web.

The other great unBLUR perpetrated over the past century is that there's a distinction between our work life and, let's call it, our life-life. The basic assumption in play here is that work is somehow not life. This is absurd; people do not become less alive or less themselves when they go to work. Again, it's the dominance of the employer in our conceptions of economic activity that gets us into trouble. It's true that time spent in the office is not time spent at home. But the economic activity of an individual is hardly limited to what he or she does in the office. And don't the economic choices people make in one sphere affect those in the other?

People simply don't exist separate from their economic selves. A woman may pay more for a certain brand of hair care products, for example, not simply for the functionality but for the way it makes her feel ("I'm worth it"). Another may choose to enter the workforce when she has the option of staying home with the kids (sometimes even at greater economic cost, given the price of day care) because it adds another valuable dimension to her life. One couple invites another to dinner as payback for a past invitation. Every human interaction and transaction involves an exchange on some level of value. They all exist in the same space, among the same nodes, in the same big economic web.

The implication of all this is that you as an individual don't have a work life to manage—much less for some organization to manage for you. This isn't about just a career. You have a life to manage.

Managing Your Life

Managing a life means coming to terms with the fact that life has many currencies that represent value. Remember the six-lane highway from Chapter 3, where we described the transaction as having three components: economic, information, and intangible value? Without always placing strict monetary value on them, people recognize many things to be desirable, scarce, and valuable. Time is a good example. A good friend is another. Intellectual stimulation, emotional engagement, respectful attention . . . the list goes on.

Don't worry, we're not about to go all self-helpy on you. This book is not titled *Women Who BLUR Too Much* (although it might be *The Man Who Mistook His Wife for a BLUR*). The point is simply that, in a well-managed life, many things are obtained and allocated with the same, if not more, care as financial assets. In fact, the choices you make regarding the allocation of your attention and emotion will have greater impact on your life than those regarding the allocation of funds. To move all this firmly back into the business book realm, networking is a great example. Everyone would agree that a great network of friends and acquaintances is an extremely valuable asset. But one of your scarcest resources as an individual is the amount of attention you have to offer, and that will become more relevant when you add the management of your own stock price to your responsibilities.

Your choices of what work to do and where to do it will also be influenced by these other, nonfinancial considerations. Work, after all, is not just about making money. At various times and for various reasons, you will require other forms of payment for your efforts. The obvious example is skills and knowledge building; you'll accept lower pay for one assignment than another if the opportunity to learn something important is part of the deal. This is why entry-level employees get paid less than experienced ones; essentially, they're still paying for an education. Other factors you might value in lieu of higher salary include enabling infrastructure, better access to people you want in your network, a better platform for your personal renown, more flexibility in work scheduling, more creative work content, and others. All of this goes into the exchange.

All this means that our work lives ("career" is a very unblurred way to think about these issues) will be indistinguishable from our life lives. Each will be as rich, varied, and unique as the life to which it contributes. Inevitably, this also means work will be as resistant to planning as life is. The best piece of advice on how to succeed in business: Don't plan too much. If you try to look across the web and forecast your position in 10 years' time, chances are that part of the web won't even be there when you arrive. It's far wiser to adapt as the information around and ahead of you changes.

MANAGING IN THE WORLD OF FREE AGENTS

We hope that so far, you've been reading this chapter with your own life and work goals in mind—free agency, direct market access, work/life integration. Sounds great, doesn't it? Now put on your manager's hat and take another look. Be afraid. Be very afraid.

From the standpoint of an organization that must draw on the labor market to accomplish its goals, increased fluidity and churn in the ranks mean it will be more difficult than ever to retain your most valuable contributors. The fact that individuals are more concerned with augmenting their skills and networks as a way to increase their personal stock price means you'll have to invest much more in development and mentoring. The prospect of individual securitization and market valuation will bring strange new dynamics to salary discussions. Simply put, the organization is not in the driver's seat anymore. It has become an offerer of opportunities rather than a buyer of labor hours.

Recruiting and Retaining Talent

One point that is abundantly clear already is that businesses will have to go to increasing lengths to attract good people. For many companies, including practically every start-up, this is the single biggest competitive issue. Vying with much larger, richer competitors, small employers find they simply can't match the salaries being offered to highly educated workers. For many, the alternative has been stock options, which offer a potentially tremendous financial upside and have the additional appeal of being attractive to exactly the type of person young companies most need: smart, boldly entrepreneurial overachievers. Using this approach, Microsoft created more than 21,000 employee millionaires; in 1997 the *average* employee's stock options increased in value by over $1 million. This, of course, relates directly to the individual's own "stock price." The worker knows, "My stock behaves like a security, performing according to my skills, my relationships, and my equity in my company's stock; a mutual fund itself, really, in that it is investing in how I and my coworkers perform."

Stock options aren't the only honey that employees are spreading as bait. At Barnett Banks' headquarters in Jacksonville, Florida, management arranged with the local school district to place an elementary school on the premises, specifically for the children of its hundreds of employees.[5] The move was doubly smart: Not only does it make people want to work there, but it lessens the chance that they'll want to leave. Companies support their employees' busy lives in other ways. It's becoming commonplace for offices to subsidize on-site services such as dry cleaning, take-home meal preparation, travel planning, and concierge staff for the exclusive use of their employees.

More valuable than any of these perks, though, is the work itself. People like the power of a strong infrastructure on which to develop and leverage their talents. Infrastructure is what keeps many a skilled and highly marketable professional in thrall to an organization that pays him less than half his billing rate. Does anybody really enjoy getting down in the trenches and adding toner to the photocopier or updating the mailing lists? And the prospect of trying to do your own computer maintenance is just too hideous to consider. (That's why these services were productized back in Chapter 2.) Far better to be able to focus on your core competence and on the new skill areas that will make you more marketable.

To be a true differentiator among companies, infrastructure has to go further than ever—and in many directions because one person will be smitten by leading-edge technology, another by a publicity department with the power to make her famous (which is also good for the stock price). Many will be lured by the leverage promised by proprietary knowledge bases and best-practice work processes. And an infrastructure that supports telecommuting and other flexible work arrangements will, for an increasing number of workers, trump everything. The key is this: A firm must attract talent and talent must maximize its stock price. Both happen when a company provides its talent with the infrastructure, support, and other elements that allow it to add the most value. For the company, managing the infrastructure is what managing stock price is for the individual: Do that, and most everything else will take care of itself.

Developing Potential

In a climate of constant employee churn, some companies are shying away from investing in training. Why put money in a resource only to have it jump ship into the service of your competitor the following month? The answer: It isn't a sustainable course to ignore your people's development needs. You'll only lose them faster that way—or never attract them in the first place. The keys are to maximize the quantity and quality of development done on the job, and to productize this service so you do it better, faster, and cheaper than anyone else.

It's also important to realize that the whole nature of development is changing in the world of BLUR. What does it mean to develop people in a labor market focused on free agency, individual securitization, and work/life coherence? One important thing you can do is help upgrade their relationship portfolios. Perhaps there is a person or organization that was especially valuable to you earlier in your career. Pass the contact along to someone who can use it now. While you're at it, introduce her to other sources of work, which may allow her to develop skills she wants that aren't particularly valuable to you. Although this is counterintuitive, because it's clearly not encouraging people to devote all their working hours to you, it's making the best of what you can't change, anyway. (Remember all that Connectivity?) If you're the one who puts an employee on the path of satisfying work—ideally with an organization that is a key part of your economic web—you avoid losing her to a competitor and maintain that "resource" in your web.

Ultimately, the best way to invest in your people will be to do so literally. As individuals more frequently choose to go public and sell financial offerings securitized by their future income streams, it will be possible to buy into them. For many, the offering will be with development needs in mind (raising money, for instance, to sustain them while they spend two years full-time working on a graduate degree). It's common to require new executives to buy stock in the company as a guarantor of commitment to performance. Why not require the company to buy stock in the individual?

Another example of how to satisfy development needs comes from a high-tech executive who reported being "interviewed" by a job candidate with rare, valuable skills in artificial intelligence software. The candidate, who was just finishing school, demanded that all benefits be fully vested and portable since he fully expected to move on in a couple of years. The company agreed, and hired him (more properly, they agreed to the exchange).

Leveraging Churn

The world's biggest public accounting firms have long understood something that other companies will have to recognize today and in the future: that most of their former employees, if anything, gain importance as they walk out the door. It's taken for granted in the Big Six—once the Big Eight, and soon the Big Four or less—that among them they'll skim the cream of new accounting graduates. This is because those graduates recognize the firms as their ticket to fortune. If they put in the requisite time on nit-picking assignments and brutal busy seasons, they can cash in big time. Turnover, given this pattern, is tremendous.

But accounting firms have succeeded in spinning a problem into an opportunity, and give their exiting staff extraordinary support. No hard feelings—far from it! Amazing as it first sounds, they know their people are more valuable after they've left. Can you imagine listing something that has vanished as an asset? Well, the point is, they haven't vanished. After all, where's the person going? Almost certainly to the internal audit or management staff of a client or potential client, where he will be in a position to influence auditor selections in the future. Second, these companies all have very active alumni programs. They keep track of former employees and give financial support to newsletters with lively "whatever happened to" sections.

Some consulting firms and investment banks follow similar policies, and for the same reasons. The famous "up or out" career path policy at McKinsey and Co., for example, also serves as a job placement service. Out often means in as a top executive with a client company, preferably in a role that includes responsibilities, and a budget, for hiring consultants.

What these firms are doing is kids' stuff compared to how you'll be managing in the near future. Remember: As churn becomes a fact of life, it won't be enough to weather it. You have to learn to exploit it. Most of the people who come through your doors will not remain in your direct employ but in your economic web for a long time.

Good compensation awaits the manager who can juggle all the requirements of the job in a world of free agents. These will be network managers, the kind who can recognize and attract talent, architect differentiating infrastructures, optimize attrition—and take responsibility for their own words when they come back to haunt them. After all, haven't you been telling your employees for years that they have to behave like entrepreneurs? Every day, you tell them, they should ask themselves: "Am I adding value to the business?" Well, from now on, prepare for them to ask such questions as, "Who gets the value I'm adding?" and, "Is this business adding value to me?"

May the BLUR Be with You

What are you doing with your attention? Are you investing it in a way the market can value? Are you listening to market signals and adapting? As a manager, are you ready for your workers thinking and acting this way? Are you ready to help them securitize themselves?

CHAPTER 7

CAPITAL

Possession Is Nine-Tenths of Yesterday

Accumulating productive capacity has always been the means by which economies grow, from seed corn to factories to mutual funds. Now the focus is shifting to your knowledge and relationships. Capital, too, is connecting, picking up speed, and becoming intangible. As it does, its future capability to create value becomes far more important than its cost. Productive capacity will be bought and sold at auction, rather than built on a balance sheet. And the most precious productive resource isn't even connected yet: attention.

In the 1970s, Citicorp Chairman Walter Wriston perceived that "information about money is more important than money itself." This comment was an early insight into Intangibles. The bank built a network that has the capability to transfer funds instantaneously around the world so it can use its assets anytime, anyplace. The bankers close up shop and go home every evening, but the money stays out all night.

When @Home went public in 1997, its shares fetched such a high price that they added up to a market capitalization of $800 million—this despite the fact that the firm's book value (the value of its asset base as reported on the balance sheet) was only $7.6 million of shareholder equity. "Irrational exuberance"? The investors didn't think so. For them, the irrationality was in the balance sheet, which counts irrelevant factors such as raw materials inventories and ignores the intangible assets that really drive the firm's future.

In many businesses today, the single most critical piece of capital equipment is the networked personal computer. More than any drill press or stamping machine, it enables workers to be productive and add value. So why is it that no company in its right mind wants to own one? Because constant evolution in computing power means almost instant obsolescence. Unlike any critical capital equipment in the past, PCs are subject to constant churn.

Six Degrees is a company that aims to provide you a return on your intangible capital—your relationships. Invest the relationships you have in them, and they'll reward you with the new ones you prize.

We've all heard the statement that any two humans on the planet are linked by no more than six degrees of separation. Now there's an Internet-based company called Six Degrees that promises to help codify and exploit that phenomenon. When you join, you enter the names of all your acquaintances—call it your schmoozing capital. As a member, if you're hoping to meet, say, Maya Angelou, you just type in her name. Assuming some other subscriber knows her, the software finds the friend of a friend of a friend who could make the introduction.

We've said it before: The work of an economy—and the firms within it—is to draw on available resources to fulfill market needs. As your firm takes account of its available resources, of what it has to work with, it's looking at two fundamental elements: people and capital. In the last chapter, we saw how the human resource is changing in a world marked by Connectivity, Speed, and Intangibles. Here, we're arguing that just like everything in the BLUR economy, capital ain't what it used to be.

What Is Capital?

To understand what's blurring in capital, some definitions and historical perspective are in order. Strictly speaking, the term *capital* has always referred to the accumulation of productive capacity. In other words, it's the enduring infrastructure that must be built, acquired, pulled together, and maintained to support production of a stream of goods or services for consumption. Most of us, when we think of capital, think of two things. First, we envision factories, machines, warehouses—all the stuff Karl Marx called the means of production. Second, we think of money, the financial backing required to set up those capabilities.

The earliest form of capital would have been purely of the physical nature, stone axes in the Neolithic economy, or seed corn in the agrarian economy, for example; in other words, capital goods and not financial capital. Prior to the early feudal times, the Western world had only so-called subsistence economies, which were hardly economies at all. There was bartering—I might give you one of my pigs in return for so many days of your carpentry services—but such activity produced no surplus in the system. Sure, a great harvest one season might produce a surplus, but whatever couldn't be eaten would only rot in the next. There was no way to convert what was left into enduring form. Without surplus, there could be none of the accumulation that constitutes productive capacity.

Feudal economies represented an early arrangement that allowed surplus to be extracted from the activities of laborers, then accumulated. The key was that feudal lords had a way to convert the surplus into durable form: They took the tithes they received from serfs and used those goods to pay the laborers, who otherwise would have been at work in the fields, to build their castles and cathedrals.[1] Those monumental edifices were the accumulation of surplus value, and became the capital that neighboring lords most wanted to grab as a way to expand their own domains.

It wasn't until we entered what is known as the craft and guild economy that financial capital became a pronounced feature of the economic world. At that point, the level of labor specialization had reached a point at which simple bartering could not supply a family with the goods it required. Striking the complex three- and four-way deals that would convert horseshoes smithed in the morning into chicken on the table that night would have taken all day. Coins were struck as convenient tokens of value storage—another, more fungible way to accumulate economic surplus.

This was a fortunate innovation, for when the technologies of the Industrial Revolution came along, vast amounts of capital were required. It takes more than a village's tithes to build a Bessemer converter or a catalytic cracker or an assembly line. This is why the so-called robber barons of the nineteenth and twentieth centuries—the Carnegies, Morgans, Mellons, and Rockefellers, for instance—were as tied up in banking as they were in railroads, steel mills, and oil fields.

By the end of the Industrial Revolution, when factories and machines were the embodiment of capital, the financial system had grown into a full-size shadow of the world of physical capital. No longer viewed as a token of value, money had value unto itself. People perceived that the accumulation of surplus did not have to mean converting those tokens into physical capital. They could accumulate surplus themselves and place it into storage facilities like banks or stocks. With the development of banking, they could even borrow against accumulated surplus. Financial institutions like banks and insurance companies began to proliferate, and financial capital, the shadow or derivative, took on a life of its own.

This is where we are now—almost. Interestingly, financial capital is such a presence today that it casts its own substantial shadow, a derivative of a derivative. The system of information that supports financial transactions has now matured to the point that it's seen by many as an even better receptacle of accumulated surplus value. This is the implication of Walter Wriston's statement at the opening of this chapter.

Perception is the key. Capital continues to change its appearance, or guise, and has come to mean different things to different people at different times. Wriston's view made him an early inhabitant of the blurred world. Yes, capital is a constant, in that it represents an accumulation of productive capacity. But physical and financial capital now have company as "intellectual capital"[2] begins to emerge. It, too, represents the accumulation of productive capacity, but the realization behind calling anything intellectual capital is to acknowledge that our most valuable assets, and our most enduring means of production, are knowledge, talent, and experience.

Stocks and Flows

Let's think again about the concepts of capital we've all grown up with. Spend a moment calling to mind something that would traditionally be thought of as capital—you know, something a robber baron would own. First, it's something enduring; unlike a consumption good, it has a long useful life. Second, it's something solid. It stays put. It's easy for auditors to check out. And third, it's something distinct, monolithic, even. It has value as a stand-alone asset, and is the property of one enterprise. Hmmm. Slow, tangible, and unconnected. Doesn't sound like something that will survive in the new atmosphere of BLUR!

What remains is that capital refers to economic capabilities to make something else, not something that fulfills an individual's desire directly. In a world of Connectivity, Speed, and Intangibles, the form that such productive capacity and accumulated surplus takes is changing dramatically. After all, what does it take to go into the shoe business these days? Certainly not a shoe factory. No raw materials, and no fleet of delivery trucks. Nike became the leader in its industry by keeping all that kind of traditional capital off its balance sheet, putting it in the hands of its suppliers instead. Nike's own value-producing capacity is its design capabilities, marketing acumen, positioning, and distribution channels. Together, these accumulate into intangible strengths that yield much higher returns than would traditional capital investments. The company's biggest physical assets may well be its spokespeople—and its contracts with them, which represent options on the securitization of future image. Talk about Intangibles! Nike has also achieved impressive mileage out of its "Just Do It!" slogan, though that is obviously something of an oxymoron for an operation that has thrived by practicing the exact opposite: "Just don't do it!"

Other companies are following the same strategy, notably Sara Lee. The company has some of the best known brands in the world, including its eponymous cake mixes and other foods, Wonderbra, Champion sportswear, and Hanes underwear. The company is planning to get out of manufacturing clothes entirely and instead will attach its brand names to goods made by others, just as Nike does; it is also considering when to sell its food business. Some companies started off this way. Britain's Advanced RISC Machines, or ARM, never did have a product other than a chip architecture that gets a bigger bang out of small batteries. ARM sells this technology to makers of hand-held computers, mobile phones, and the like.

A few big ideas encapsulated in these examples deserve to be parsed out, because they apply to every enterprise today.

MOBILE CAPITAL

First, capital in the traditional sense is no longer the basis of enterprise value. Competitive success simply doesn't accrue to those who are sitting Yertle the Turtlelike on the biggest pile of assets. Quite the contrary; in an economy marked by unprecedented Speed, what's valuable is not what's standing still, but what's in motion. Perhaps we need to walk away from the idea that owning or even controlling capital is a necessary resource for fulfilling market needs. More likely, we need to learn to use capital as a way to measure productive capacity as a flow and not a stock. Either way, the focus of value must shift from the static to the moving.

How much productive capacity—ideas, skills, innovations—
is your business flowing through its systems every week?

JUST-IN-TIME CAPITAL

The second general lesson is that it often doesn't pay to own capital
equipment. In a fast-changing business environment, there's great
risk in ownership. Chances are (a), that the equipment will become
obsolete years before it's worn out. You don't count all those out-of-
style clothes in your closet as capital, nor depreciate them as styles
change, so they're not capital. Neither is your two-year-old PC, even
though it works as well as it did when you bought it. And (b), ownership can prove to be an albatross, the sheer weight of which will
hinder the firm's ability to move swiftly out of one business line and
into another. Meanwhile, the risks inherent in not owning one's
own capital equipment are diminishing rapidly. The potential for
forging and maintaining tight links with suppliers and outsourcers
grows greater every day, and the chances of being left high and dry
by them are minimal and manageable. In other words, the new
economy's Speed and Connectivity argue against physical capital
investments. Capital as inventory of capacity must give way to "just-
in-time" capital as access to the use of capacity.

How much capital does your business own? Would you be
better served by letting someone else provide some of these
capabilities?

INTANGIBLE CAPITAL

A third and final lesson from Nike (for immediate purposes) is that there are other, more valuable forms of capital than factories, machines, or even money. What constitutes productive capacity today? In an increasing number of businesses, the real capital is intangible; it consists of such things as brand image, strong customer relationships, the talent on staff, the experience built into business processes and systems, and so on. These are the elements that are best leveraged into additional profitable production. The implication is clear: In our future accumulations of surplus value and investments in production capacity, we have to get less physical. The tangible must give way to the intangible.

These are the BLURs we are seeing in the world of capital. In the following sections, we'll devote more time to them as we explore further the blurring of stock and flow, of production goods and consumption goods, and of hard assets into Intangibles.

Value What's Moving, Not What's Standing Still

What becomes a corporate legend most? It used to be that a solid balance sheet was the most accepted indicator of enterprise value. The measure of a company's greatness was its asset base, made up of all the land, buildings, equipment, and inventory it owned outright. The balance sheet added up all that property, alongside all existing claims against it. Like a snapshot, it portrayed the position of the business accurately and in a way that all interested parties could easily comprehend.

But, like a snapshot, the balance sheet reported on just one point in time. It told nothing about trends upward or downward; it gave no context in which to assess future prospects. In many cases, it painted a much more sanguine picture than was really the case; a company's top accounts might be defecting left and right, its workforce might be out on strike, but still the company's value looked intact. Just as often, companies were short-sheeted by their balance sheets. A firm with a groundbreaking product might be experiencing brisk, high-margin sales, but thanks to debt taken on during research and development, might look like it was nearly bust. Imagine you owned the goose that laid golden eggs. The balance sheet would report holdings of one goose, but it would assign no value to future expectations, and thus miss the point.

It has been only in the past half-century that most businesspeople have realized it's more useful to focus on flow and change rather than on stock and stasis. With this realization came the rise of the income statement as an indicator of enterprise value. The income statement, of course, is a sort of elaborate footnote showing how the company arrived at the "retained earnings" line; a "stock" concept, not a flow, in the balance sheet. Whereas the balance sheet shows balances as of a specific date, the income statement shows the flow of activity and transactions over a specific period of time. As such, it depicts in a much richer way the true health of the business. Future accumulation of value, it becomes clear, is more closely related to income than to assets. As a consequence, debt began to rise as a proportion of equity, because a company was not priced for its net worth—a measure of the capital it owned, free and clear—but for the income it could generate. Return on equity replaced return on assets as the dominant measure.[3]

Now the focus is shifting again. People analyzing enterprise value are more concerned with future prospects than current performance. The change is most apparent on Wall Street, where the valuations of firms are reflected in their share prices. No one has missed the fact that those prices are yielding market capitalizations for many firms that far exceed their book values. On average, the market capitalization of a company quoted on the New York Stock Exchange is two and a half times its book value. An investor simply wouldn't arrive at those values by focusing on the flow reported in the income statement (much less the stock reported on the balance sheet). Is Wall Street suffering another attack of irrational exuberance? No. It's just that when it looks to determine value, the Street doesn't focus on assets, nor does it focus on income. Instead, investors search primarily for a promise of growth.

This chart summarizes how the methods of valuing a company have shifted over time, attempting to further clarify the shift through analogies from physics and mathematics. Again, as businesspeople attempt to assign a value to a company, they can focus on its assets, its income, or its growth rate. These are, respectively, expressions of

THE BLUR OF CAPITAL

Type of Capital	Valuation Focus	Who Focuses on It	Economic Concept	Physical Analog	Mathematical Analog
Physical	Assets	Auditors	Stock	Position	Number
Financial	Income	Analysts	Flow	Speed	First Derivative
Intellectual	Growth	Venture Capitalists, Simulators	Acceleration	Acceleration	Second Derivative

stock, flow, and acceleration. By analogy, picture yourself at the stock car races placing a side bet after the gun has gone off. You could look at the relative positions of the cars at the moment. Probably this is not too useful, since it's still early, and the one who was in pole position has a deceptive advantage. Better to focus on the relative Speed of the cars at the moment you placed the bet. But, if you could gauge it, the really useful factor to know would be their relative rates of acceleration. It's the car that's picking up additional Speed at the fastest rate that's most likely to win. In mathematical terms, the shift is from focusing on a simple number to focusing on a derivative of that number, and then to focusing on a second derivative of the original number.

Yugi Ijiri of Carnegie Mellon recognized this shift early; for the past decade, he has been refining a proposed addition to the standard set of financial statements. The purpose of the addition is to depict the momentum of change in the company (that second derivative). He refers to the system he's constructing as "triple-entry bookkeeping" because his acceleration statement would inform the income statement as much as the income statement informs the balance sheet.[4]

GETTING CAPITAL IN GEAR

What does all this mean for capital in the blurred world? A few pages ago, when we were telling you to hollow out your business and use the capital of others, you probably wondered what to do if you choose to be the one holding the bag. First, accept that capital in the traditional sense is a false god. It doesn't make sense to stockpile great amounts of physical assets for their own sake. They have little intrinsic worth. We've often heard that a human body, as valued by the sum of its elements—mainly water with particles of a few semi-valuable chemicals—is worth about $2. The same applies to companies: Asset ownership bestows little credit; it's what you do with them that counts. At rest, they rust or decay, and ultimately end, as does a human body, as dust. Put those assets in motion, however, and you create living value.

Second, and more important, this means that every bit of capital you do own should be kept not only in constant motion but at an accelerating pace. The faster capital works, the less of it you need. The point applies to physical capital as well as to financial capital. It's only in systems where inventories sit unproductively on shelves that everyone needs a lot of them. This means, for instance, that if you have the ability to cut and sew leather, you shouldn't become a captive resource of a Nike, but rather should create an economic web of your own for New Balance, the Tannery, Hermès—whomever—so that you can keep your capital equipment in constant motion (connection) and learn from each of the businesses you serve (intangibles).

When matter accelerates toward the Speed of light, it picks up mass ($E/c^2 = m$). In the same way, as income goes around faster and faster, at some point it takes on the economic weight of capital. Hence, you can borrow on the promise of an accelerating income stream as Marimba, Mainspring, and countless other Internet start-ups have done. That's why, if you own the David Bowie bonds referred to in the previous chapter, you'll be able to leave a chunk of his income stream to your heirs. *When you decide to own and manage capital, value what's moving, not what's standing still.*

The Blurring of Capital Goods and Consumption Goods

The easiest way to understand the difference between capital goods and consumption goods is to understand their yin and yang: Capital goods are the physical capacity that enables the production of goods for consumption. Traditionally, it was easy to tell the two apart because consumption goods were ephemeral, while capital goods endured. The detergent you bought at the store was used up in a month; the vats used to make it were in place for decades. The notion of durability, in fact, is built into the accounting definition of capital. In order for a purchased good to be "capitalized" on the balance sheet, it has to stick around for at least a year.

This reliable old distinction, however, is becoming more subtle. Given the Speed with which businesses and product lines change—and, yes, blur—it's clear that, whatever time span "enduring" used to mean, it's a lot shorter now. Last year's machine tooling is likely to be no more marketable than last year's fashion, and as useless as last year's bread. Enduring is also a problematic concept when so much of today's productive capacity is made up of intangible assets. The programmable controls now featured on assembly-line equipment are a good example; they muddy the question of how much of it is enduring and how much evanescent. The hardware is lasting and becoming less valuable than the short-lived and more valuable Intangible: software.

Finally, adding Connectivity poses its own challenges to the distinction between capital and consumption goods. When the chip in the car remains connected to the computer at the factory, where does the capital good leave off and the consumption good begin?

The point here is that capital goods are more frequently behaving like consumption goods. One implication of this is that any acquisitions should be weighed carefully, with an expectation that obsolescence will begin to set in immediately (i.e., focus on acceleration, not Speed). Even more strongly and surprisingly, it means we should actively work to *minimize* the lifetime of capital goods. In other words, get as much out of them as possible while the getting's good.

This basic advice translates into four axioms about capital in the blurred world:

1. Use it, don't own it.
2. If you do own it, use it up.
3. Design to throw away.
4. Design to reconfigure.

Let's take at look at them in detail.

USE IT, DON'T OWN IT

"Possession is nine-tenths of yesterday" is another way of saying that the best way to avoid the perils of capital obsolescence is not to own it in the first place. Of course, this is the sentiment that is fueling the rapid growth of the outsourcing business. Outsourcing simply means sending out work that used to be done in-house, and it makes sense in any area that is capital-intensive but not truly core to the business (or not needed often enough to justify full-time staff). For a while now, it has been common for manufacturers to outsource their shipping activities. Other functions routinely outsourced include research, advertising, distribution, maintenance, and payroll processing. The New England, a Boston insurance firm, decided in 1992 to clean up its balance sheet by outsourcing its entire marketing function. It did so by spinning off the old department into a separate entity, which then had The New England as its biggest, but not its only, client.

Where outsourcing has caught fire in the past several years is in the realm of information systems management. Here, the motivation is more interesting and more BLUR-related. Few firms, after all, would claim that their information is noncritical to their business; for many, relinquishing control of business information is akin to handing over the crown jewels. Most would also laugh at the suggestion that IS management is not a full-time need. But outsourcing still makes sense, first, because of the capital-intensity of information management, and second, because of the outrageously quick obsolescence of the computers involved. The cost of having to churn equipment is punishing—and it isn't even the worst problem. The really hard part is the people. Rapid evolution in the field makes their skills as obsolete as their equipment and introduces new risks to information integrity and security. Besides, what IS superstar wants to work for a firm where IS is only a support function?

Connectivity in the economy, of course, is what makes effective outsourcing of IS management—and other nontrivial functions—possible and preferable. It's also what will enable more activities, and all the capital that goes with them, to be removed from both the company's premises and its balance sheet. Such moves are hardly limited to the world of computers since management's first impulse is to outsource any kind of work. In the blurred world, the rule will become: Don't own anything, unless it is absolutely your core competence. Even then, consuming it as rapidly as possible will be the way to success. It may still be true that the physical plant is the anchor of the enterprise, but anchors aren't what you need when the goal is to move forward fast.

IF YOU DO OWN IT, USE IT UP

Now let's look at the other side of the outsourcing coin. For those who do choose to be in businesses that necessitate major capital investments, the name of the game is to squeeze all the value possible out of them before they become obsolete. Of course, working your capital hard is nothing new. It's why McDonald's started serving breakfast after decades of offering only hamburger lunches and dinners. But the issue isn't just about using hard assets over long hours each day. Rather, it's about recognizing how soon a new technology will come along that makes the equipment you now own less valuable or, even, a liability.

All this, of course, explains the strategy of companies like EDS and Perot Systems, both of which specialize in business outsourced by others. In three years' time, the computers they own will be useful only as paperweights; the goal is to utilize them more than anybody else would by keeping the drives spinning 24 hours a day. That means selling to all comers until the last bit of capacity is spoken for. Just as important, they must excel at estimating the trading price of the equipment three years down the road since any profit may well depend on the residual value of those hard-working, outdated computers.

The same economics drive the printing business. This is an industry that has been completely reinvented by technology in the last decade, and the pace of change just keeps accelerating. Presses are still built to last, and a decent one will keep going for 20 years;

but three years down the road, its shortcomings will already be obvious. In the meantime, keep it rolling.

All capital equipment is becoming more like real estate, in that while the building remains structurally sound, a downturn in the neighborhood can all but wipe you out. Think about the investment it takes to make the latest memory chip. When 32 megabits is the biggest, the rewards are high—but only until the 64-meg chip moves in across town. There goes your zip code. Your address tumbles from a tony computer community to a decaying middle-class neighborhood. You have to redeploy your now obsolete, stepper technology to a jelly bean that lets greeting cards talk.

Let's say you hold the patent for the hottest thing since Cabbage Patch dolls. Chances are very high that your Cabbage will be as cold as coleslaw in a year's time. The last thing you should do is keep it to yourself. To wring all the value possible out of it, strike licensing deals left and right. This is what Investcorp did when it bought and revived the faltering Gucci fashion accessory business. It saw a strong brand capable of carrying a much wider product line than the original company produced. Now, there are Gucci watches and perfumes to go with those famous scarves and loafers. Make no mistake, brands very definitely are capital. They represent accumulated surplus value turned into loyalty, which translates into lower marketing costs, higher prices, or larger market share.

What we're saying—invest in expensive capital equipment, turn it around, trash it—would doubtless delight a Luddite. And yes, it sounds irresponsible, even cynical. But the fact is, there isn't an option; capital today has to live fast and die young. So change your perspective and get over it. If the candle will outlast the night anyway, then by all means burn it at both ends—and enjoy all that lovely light while you can.

DESIGN TO THROW AWAY

The clothes-horses among us have always known not to bother having our high-fashion threads too well tailored. If it's going to be out of style in six months anyway, why pay the price of durability? The same sentiment is now driving the manufacture of other consumption goods—watches, cameras, and telephones, all of which were formerly classified as consumer durables—and will just as surely shape the capital goods of the future.

In some cases, it will be the capital equipment itself that is destined for quick obsolescence; in other cases, it will be the consumption goods the equipment is designed to produce.

DESIGN TO RECONFIGURE

The best capital investments will be in equipment and capabilities that prove robust enough to support a changing product line. To the extent that it's possible to see the direction that future offerings will take, it makes sense to design means of production that anticipate the change. Auto assembly lines, of course, do this to a large extent. It would be crazy to have to reconfigure and retool the factory floor with each new model year. Other than a given number of "hard points" that can't be changed, the line is designed to accommodate new features without adding major new costs. Of course, as software represents more of a car's value—both the software to support flexible production and the software built into all the systems of the car—this gets a lot easier.

In the computer industry, designing for reconfigurability is absolutely essential to competitiveness. With product life cycles measured in months, technology firms simply don't have time to overhaul their infrastructure between product introductions. The rage now is to establish "product platforms," which simultaneously prepare for a major product innovation and for the next several derivative products it will spawn. Platforms work not just for computers and automobiles; every investment in design can be seen as a platform. Airplanes have been designed this way for a long time. Boeing's 747, for example, was a platform for the stretch, cargo, megatop, extended service, and other versions of what basically was the same plane. Seiko designed watch factories to produce huge numbers of short-lived styles—all based on only a handful of basic styles. George Lucas might not have known that he had a platform when he created Star Wars, but today the accumulated value of the shared vision of the Federation is managed just like that Boeing plane—it's a platform with "stretch" regions, international relations, and a long-range market plan.

Back on Earth, office space is the latest thing being targeted widely for reconfigurability. What started with prefab cubicles has migrated to entire movable walls, rolling files, and wireless networks. The trend toward "hoteling" is one aspect of this: It assigns spaces to people according to their needs. From day to day, office neighbors change. At the same time, new work styles demand more flexible space. Many firms, hoping to foster teamwork, create space that can support focused, private tasks, but that can also adjust in a moment to accommodate large-group brainstorms and celebrations. Remember what we told you in Chapter 2 about downloading your upgrades in order to keep your offer fresh? This applies to capital as well, though it's easier in some cases than others. Software capital—production equipment—is relatively simple to upgrade, of course. You merely download the new release of C++ or Visual Basic, the development environments that represent capital in that industry. In the hard goods world, however, flexible manufacturing systems will have to become considerably more flexible.

Before we leave this subject, an important note: People are a platform, too. What they know and what they can do is like a current set of tooling. The implication is that their knowledge must be fully used before it becomes obsolete, and refreshed constantly if the platform is to retain its productive power.

The Blurring of Hard Assets into Intangibles

Early in this chapter, we made a wild claim that capital was no longer the basis of enterprise value. We even took a cheap shot at Yertle the Turtle. Now we're going to temper that statement somewhat. The truth is that capital, in the traditional sense of plants, production equipment, and financing, is not such a big thing. But capital in the new forms it is taking has become more fundamental to enterprise value than ever.

What are the new forms of capital? In the past year, we've been hearing a lot about three in particular: intellectual, human, and structural. Briefly, intellectual capital is the brain power of the organization, codified and put into explicit, transferable form like a piece of software or a document. Human capital is the value of workers' relationships (with customers or experts, for example) and their tacit knowledge—the accumulated but unarticulated experience that guides both large and small choices. Structural capital is the experience and expertise of the organization embedded in processes, policies, and systems. All Intangible, and difficult to measure, but can you deny these are the real engines of profit in today's economy?

Still, to call something capital is not just to say it's important. Remember that capital is accumulated productive capacity, and that it began with a purely physical infrastructure alongside which financial mechanisms rose as a shadow system. When did cash come into its own as a form of capital? For practical purposes, it was when institutions devoted to it appeared on the landscape: the banks and insurance companies who made it their business to measure, manage, and provide it. Likewise, as information began to be regarded widely as a form of capital, it was no coincidence that firms like Dun & Bradstreet and the granddaddy of information gathering, Reuters, arose to capture it and make it available. The next step is only logical: We're seeing new institutions and new mechanisms making a business out of managing human, intellectual, and structural capital.

INTELLECTUAL CAPITAL

The hoariest of the three has to be intellectual capital, which has been a buzz phrase in business circles for a few years now. Like any great form of capital, it has its apologists. Some of the leading ones include authors Ikujiro Nonaka, Hirotaka Takeuchi, and Takeuchi Nokada (*The Knowledge-Creating Company*), James Brian Quinn (*Intelligent Enterprise*), Tom Stewart (*Intellectual Capital*), Tom Davenport (*Working Knowledge*), and Karl Sveiby (*Managing Knowhow*). All have important and valuable insights, if a variety of different specific definitions, to offer on the role of knowledge in business. What they agree on most is the basic notion that the competitiveness of a firm is more than anything a function of what it knows, how it uses what it knows, and how fast it can know something new.

Certainly, there is no lack of institutions devoted to capturing, managing, and providing intellectual capital. On the most straightforward basis, this has always been the domain of legal firms focused on patent law. The patent, after all, is the most clearly codified and measurable form of intellectual capital. More broadly, consulting firms exist purely as receptacles of specialized knowledge for sale. It's no surprise that they are also the leaders in the use of innovative technologies and management approaches designed for internal knowledge transfer. Currently, this form of capital is undergoing a change from a craft production approach—in which skilled individuals are the primary source of value—to a mass production approach, in which the skills are embedded in tooling. In the nineteenth century, dies, molds, jigs, and eventually transfer lines were the form of the embedded intellectual capital. Now, firms like SAP, Oracle, and Baan are producing the new form of tooling—so-called Enterprise Reengineering Packages—that allow mass use of intellectual capital. Some firms have offerings that recall more traditional information services but add a level of richness that makes them knowledge-based. Teltech, as mentioned earlier, is a firm that arms engineers and scientists with state-of-the-art technical awareness. A typical client is R.W. Johnson's Pharmaceutical Research Institute, where pharmacologists labor to discover new drugs. No simple broadcast service, Teltech responds to unique inquiries with one-to-one access to experts. Knowledge is delivered just in time, not just in case.

One job that has arisen out of this kind of BLUR is that of a self-employed student researcher. Reva Basch, a trained librarian, earns her living at home, searching for answers on the Net to questions asked by clients to whom she will never speak. These have included subjects as varied as male-female differences in color perception, the sunglass market, current research on Alzheimer's, and the first record-ed use of the term "guerrilla marketing." "Unless the phone rings, I may not speak, audibly, to a soul all day," she says. But she has the ability to "find in minutes what might take [someone else] hours, or even days, in the library, assuming it's there at all."[6] Basch files her report, receives her fee, and moves on. She epitomizes a provider of pure intellectual capital.

HUMAN CAPITAL

Human capital is a trickier subject. Here we're talking about material that's even harder to extract from people than the skills they have or facts they know. We're talking about the relationships they have with other people, and the esteem in which they are held. If you know a good salesperson or an effective leader, you know the value of human capital—and you recognize it as largely nontransferable. Much more than intellectual capital, it's bound up with emotion, cultural background, ethics—even basic physiology.

Human capital has always had its institutions: the country clubs and business-lunch restaurants where soft-selling and networking get done. But it's hardly fair to say those institutions exist to manage and provide human capital for business. A better argument can be made for industry associations and professional societies, whose meetings and conferences are overtly geared toward relationship building. The conference business in general is moving in this direction, though some have more star power than others. Annual events like The World Economic Forum held in Davos, Switzerland, are exclusive and high-priced precisely because of the crowd that will be there; the event is all about schmoozing with the most interesting and accomplished people in technology, entertainment, and design. Richard Saul Wurman, who runs such conferences himself, calls the Davos event "the dinner party I wish I could have." Think of it. People arrive, talk, and leave. Nothing has changed hands between them and yet human capital has been created.

Far and away, though, the most interesting example we've seen of a new business dedicated to human capital is Planet All, a start-up commercial venture on the World Wide Web. It requires new customers to catalog all their existing relationships. Planet All helps maintain those relationships with services such as calendar comparisons and birthday reminders. If your next business trip, for instance, treats you to a six-hour layover in Paris, it will suggest you might want to look up ole Bill, who has been there on assignment for two years. It also helps customers leverage their relationships into additional ones, identifying the friends you have in common with other customers, and shows you the pathways to new, potentially valuable acquaintances. More than anything we've seen, it indicates that relationship capital is being perceived as something that can be codified, managed, and leveraged.

STRUCTURAL CAPITAL

According to Tom Stewart, a member of the board of editors at *Fortune*, "Structural capital is knowledge that doesn't go home at night." It's also a huge grab bag of stuff, including as it does all the firm-standard business processes, systems, and policies that represent the accumulation of experience and learning by many people over many years. In a mature firm—or society—most of the structural capital predates most of the people currently on staff. Imagine coming into an organization where no one had yet figured out how the organization paid the rent, assigned the telephone numbers, and the like. It's the enabling infrastructure that captures how the company gets things done, and develops it—or, as likely, watches as it develops itself—over successive generations of workers. In the best kinds of firms, structural capital is constantly kept in view and on trial. Like any form of capital, it is subject to rapid obsolescence. But too often, it is so ingrained and unchallenged that it becomes a disabling infrastructure. The famous analogy is to paving old cow paths.

What are the new institutions of structural capital? A few obvious ones come to mind. First, we would cite the rise of benchmarking consortia like the American Society of Quality Control (ASQC) and the American Productivity and Quality Center (APQC). Both exist to help member organizations map the business processes they have in place and help them migrate closer to what is considered "best practice." As noted, there has been the recent astonishing rise of SAP AG, the German software maker. SAP software is essentially a template for excellent integration of supply chain processes like procurement, inventory management, production, and distribution. In the past, companies did this for themselves, with greatly varying degrees of success. Then, consultants offered human capital–based solutions. Now, firms that buy into SAP are buying into SAP's accumulation of great structural capital. Again, the very appearance of an institution like SAP reflects the growing recognition of a new kind of capital that can be captured, managed, and sold.

Attention Please

The thinking behind the identification of intellectual, human, and structural assets as capital has been insightful and even revolutionary. It's true that these are the real productive capacity in today's economy, and each has matured to the point at which it is possible to accumulate surplus. But, just like their predecessors, there is nothing sacred about them. As new things are recognized as scarce, valuable, and manageable, we will have still newer forms of capital to worry about investing well. Rest assured, it won't be information that runs dry. It multiplies without charge and so could hardly be said to be threatened by scarcity. What is tough to find, however, is enough time to use all that information. Meet the real, increasingly scarce resource: attention.

Think about it: How many messages do you receive across how many forms of media every day? How many tasks do you need to attend to, compared with a decade ago? On an individual level, some of your most important decisions have to do with how you choose to allocate your attention. For businesses trying to communicate with customers—and especially for those trying to reach potential new ones—attention scarcity is a critical issue. Equally, capturing a buyer's attention is the most challenging part of the sales process.

A senior vice president at one of the world's largest insurance companies told us recently of their inability to derive high financial returns from their core business. Yet, we're told, other companies are lining up to invest in the business. Similarly, Silicon Valley start-ups have to beat off would-be investors with a stick because of the growing oversupply of financial capital. To repeat: *There is more money around than there is competent attention to put it to use.* This means two things: First, returns on capital, traditionally measured by return on equity, will drift lower over time as the scarcity of financial capital declines. Second, return on attention will become as carefully measured as ROE was in its time.

Happily, aspects of the management of attention capital are actually more mature than management of other intangible assets. Primarily, this is because there are businesses devoted to it already, and acknowledged experts abound. The advertising and entertainment industries have always been focused on attention capture and use. Their importance will grow, along with the scope of their services and range of their clients. Engineers were kings when physical capital was where the value was added, and investment bankers were "masters of the universe" in the heyday of financial capital. Now, the top talent in attention businesses will come to the fore in the next decade. After all, what is a celebrity but someone who earns a premium for her ability to command attention—and redistribute it to a sponsor or product. Sometimes the product is related to the celebrity—a Michael Jackson video—sometimes not (think about football players opening car dealerships).

Meanwhile, we've already seen one small concern—again, Internet-based—stake out a business with attention capital explicitly at its core. Seth Godin founded Yoyodyne on a great idea: In a world of information overload, it's much easier to market to consumers once you've gained their permission to do so. This concept has led to a new niche, *permission marketing*. As Godin explains it, the concept works like this: A guy decides he'd like to get married. He goes to Barney's and pays $2,500 for an Armani suit, $95 for a Zegna tie—the works. Then he hits the right New York east side spots, proposing marriage to every woman he meets. After two unsuccessful months, he goes to Bergdorf Goodman and pays $3,000 for a Brioni suit, and starts over. This, says Godin, is like advertising. You spend a lot to dress up the message and then deliver it to people who don't want to hear it.

In contrast, a woman with the same target in mind but who has much more smarts might think about her knowledge of cigars, go to a cigar club, and strike a match if the right-looking guy needs a light. If he accepts, she might ask about his preference in smokes. . . . The courtship might not continue, but if it does end up with wedding bells, it's because he gave her permission to proceed.

Yoyodyne does this with both its commercial and consumer customers. It produces online games for various clients, including KPMG Peat Marwick (Ernst & Young's betrothed at this time of writing), targeting them at college students. In KPMG's case, the ads take the form of recruiting messages that the players implicitly agree to watch when they settle down to compete for prize money that can run into the thousands of dollars. Everybody's a potential winner. KPMG gets the attention of smart kids in a labor market cluttered with demands for student attention. The students get entertainment—and winks from possible employers. Another Net-based example of permission advertising is Cybergold, a company that pays people actual cash, usually a $1 at a time, to look at ads. In that case, they're paying cash for attention.

It's fun to imagine the new developments that will take place as attention capital comes into its own. Will new tools for managing it arise, as important as spreadsheets were to the world of finance? Imagine getting—imagine needing—periodic reports on where your attention was spent, the return it realized, and how much attention you gained. It's inevitable that a whole set of mechanisms and players will appear on the scene to help each of us invest our attention optimally.

Finally, in a world where it is understood that the scarcest commodity is attention, how will legal codes change? Surely it will be seen as criminal to abuse someone's attention, or otherwise waste their waking hours. Come to think of it, it's no better than stealing something so personal that it's beyond value. Don't you wish you could fine unwanted telemarketers who call at dinnertime? Think of the last time you sat in a traffic jam for miles before seeing the cause: a line of orange cones stretching half a mile, one desultory construction worker, and a flag man. At that moment, would you have been receptive to the idea of a class action suit? Legal action against time-wasters is not without precedent, actually. Straphangers the world over must marvel at the $10,000 fine levied by the Japanese government against the family of anyone who jumps in front of a bullet train. The message: However tragic, a decision to commit suicide does not carry with it the right to steal attention from thousands of others.

What's Scarcer than Money?

So, can it possibly happen? Can things like knowledge and attention in time come to be seriously accumulated as productive capacity? Will these new forms of capital, now shadow systems at best, take on a life of their own and ultimately eclipse financial capital? We're seeing the first stages of a shift already. New institutions are beginning to be formed to bring rigor and purpose to intellectual, human, structural, and attention capital. At the same time, as already noted, financial capital, long the scarcest resource, is not so scarce anymore. In conversation after conversation, we hear major investors complain they've got funds to spare, but the returns just aren't there. Returns, of course, track nicely with scarcity. They're going down because, relative to other forms of capital, cash is abundant. Ask your own brokers: They'll probably tell you the same. Better yet, ask yourself: What is it you're really short of these days? Is it money? Or is it enough time to pay attention?

> *What financial and physical capital is on your balance sheet? How fast are you using it? What's it doing for you? Might it better belong to someone else? How is your organization allocating, measuring, and valuing attention? Where could a reallocation of your time capital—attention—have the greatest impact? How could you make more attention?*

BLUR, the Economy, and You

We've limned the BLUR economy for you. Now you need to know what you can do about it. You'll invent more things than we can imagine—as we said earlier, we hope you'll become part of the community discussing how to BLUR business, and learning from one another by sending us your ideas, joining us on the Web, or visiting us in Cambridge. Meanwhile, here are **50 ways to BLUR your business . . . and 10 ways to BLUR yourself.**

PART V

Living

the

BLUR

CHAPTER 8

50 Ways to Blur Your Business and 10 Ways to Blur Yourself

What you can do differently in business, your organization, and your career to be a winner in the BLUR economy.

We've covered a lot of ways to put BLUR into your business life and/or your personal life—and let's face it, the two are blurred anyway. Here are 60 that we particularly like.

50 Ways to BLUR Your Business

1. Make Speed Your Mind-set

Everything exists in the fourth dimension—time—but few of us have a good handle on it in our businesses. To reorient your thinking, start by timing everything you do. Then decide how long it should take you to cut that time in half, then in half again, and so on until you can do that thing almost instantaneously. If you typically introduce a new product each year, for example, focus on introducing the next one in six months, and the one after that in three. Figure out how to speed up the move from concept to customer, from new service idea through its development and delivery. Then decide what you have to do to cut the time on these kinds of tasks in half, and—you've got it!—in half again. Your customer should not have to wait for service, your supplier for needed information, your partner for sign-offs, and so on. Learn to *think* in real time. Set a goal for matching this state.

2. Connect Everything with Everything

Make sure that all the islands in your business connect with each other, just as the empty shelves in your refrigerator ought to connect with your shopping list. Identify what's unconnected. For example, maybe your internal information systems just aren't hooked up to one another for maximum Speed and efficiency. More challenging, you must also connect to islands that lie beyond the boundaries of your company. Your customers need to connect with your engineers, your potential recruits with their likely future bosses. You may also need to connect with your competitors, because in the blurred world, you may well depend on them as future partners. Risky? Sure. Connect your design group to the embedded base of products you've sold. Some connections are extremely tough, but when they're part of a connected system, they all pay back more than they cost.

3. Grow Your Intangibles Faster Than Your Tangibles

Figure out which parts and what proportion of your offer are intangible (which we'll call "i") and which are tangible ("t"). Then figure out how to grow the Intangibles faster so as to increase that i/t ratio. When listing your Intangibles, include all services, all software and other forms of information, all financial elements, and all emotions (which include brand loyalty, customer relationships, employee commitment, and the like). Tangibles include actual products of the type that come off an assembly line, bricks and mortar, and anything that rusts or that you can bump into. More complicated: The people on your payroll are tangibles, while what you really value about them—their experience, intellect, and overall brain power—are intangible. The BLUR secret is to move these folks from the tangible to the intangible column. Subcontracting and outsourcing are two ways to do this—maybe even buying a position in them.

4. Build Product into Every Service

How do you offer service today? Does it meet customers' demands for speed, accessibility, quality, responsiveness, and customization? The only way you can afford to meet those expectations is to productize your service: Use software, kiosks, self-service, learning engines, or telecommunications to deliver service like a product.

5. Put Service into Every Product

In the past, service add-ons were an afterthought for those in the product business. Today, you must make them intrinsic to the offer. If your entire corporate culture is focused on *stuff*, wake up. Services often provide higher margins and better growth opportunities than the rapidly maturing thing itself. Maintenance contracts are like the Model T. They're great to look at, but relics of the past nevertheless. These days, you also need to offer extras such as 800 numbers for contacts, compliments, and complaints. Better yet, build computers or chips into old-time products. Examples: Computers in cars help you navigate and drive better. Seiko's new watch keeps you up to the minute on stock prices.

6. Manage All Business in Real Time

Stop making decisions based on what happened last week—or even this morning. Get a grip on what's happening at this instant, so the right adjustments can be made without delay. Almost always, this will require planting sensors and other feedback mechanisms throughout your operations. At Austin Quality Foods, sensors in cracker ovens gauge and report instantly on moisture levels (and therefore, doneness). Before they had the sensors, quality control staff had to run frequent analyses off-line. When there was a problem, it took a half-hour just to catch it, during which substandard product continued to roll off the line. That meant a lot of waste—all avoidable. Japan Central Railway has sensors all over its tracks, which transmit data instantly to a real-time simulation of its network. If a breakdown occurs, managers immediately route traffic accordingly. It never has the chance to create a bottleneck. Don't limit this approach to physical processes. Monitor how many recruits were interviewed, how many sales calls were made, or how many loans were sold, in real time.

7. Be Able to Do Anything You Do at Anytime

Nine to five has been dead for more than a decade. These days, 24×365 rules. Retail banking customers today would rebel if they didn't have round-the-clock access to their money via ATMs, PCs, or telephones. Getting your organization running according to the same we-never-close clock might mean deploying that same kind of automation, letting night-owl customers serve themselves. But for many businesses, it will require operations to follow the sun. Chase and other global banks used to have all night to do their batch processing. Now, it must be done for Tokyo before the business day begins in London.

8. Be Able to Do Anything, Anyplace

With the reality of today's Connectivity, you should be able to conduct your business no matter where you—or your customers are. This means that you have to set up your sales, service, and support systems so that your customers can access them wherever they may be—on the Internet, near a phone, within reach of a fax machine, or grasping their PalmPilots—anywhere in the world, on the ground or at 40,000 feet. Cisco Systems does this by putting a huge amount of user support information on its Web site—in 15 different languages. For those who must speak live to a support person, it provides anywhere advice supported by the AT&T Foreign Language Translation Services.

9. Put Your Offer Online

How online are you, really? Sure you've got a Web page, but how much of your business is ready to move once your customers click on those buttons? Having your offer "within an arm's length of desire" enabled Coca-Cola to build an empire. Today's electronic Connectivity allows the same simple strategy to work for everybody. Your customers must be able to connect with you and your offer at least as easily and freely as the Coke drinker can pick up a Coke. The Net is the first distribution channel of the electronic age. You must learn how to use it and the more robust networks to follow for advertising and exchanging (selling and buying) your business.

10. Make Your Offer Interactive

So you think you listen to your customers? You're *truly* listening only if you maintain an ongoing dialogue with them. Your offer must contain a way for your customers to communicate with you as they put your offer to use. Make an inventory of your offerings and take a close look at how interactive each is in terms of a continual dialogue. The ongoing exchange of offers grows ever more valuable with the passage of time. Make the most of it.

11. Customize Every Offer

It used to be one size fits all. Now, Porsche says it never makes the same car twice. Whatever your offer, you must tailor it each and every time, with the needs of the individual buyer or user in mind. Cheeseburgers, hotel rooms, pants, software programs, office chairs, retirement plans, skis, kitchen appliances. . . . Today, every offer, no matter its nature, can be customized, along many dimensions. Crunching and connecting tools make it possible; your need to survive makes it essential. If your product is a commodity, customize the service you wrap around it. If your service is cookie cutter, let the customers "have it their way" via individualization. (Imagine: a cookie in your child's image!) Customize, customize, customize. It is the great differentiator of the twenty-first century.

12. Make Sure Your Offer Gets Smarter with Use

Build observation capability into your offer so that it can teach itself to do things better. Banks have credit lines that increase in response to a customer's good payment record, at the same time improving the value of the bank's offer. Your offer must be able to learn from the way your customer puts it to use, and respond to the communication. The next iteration of offers you extend should incorporate that new-found knowledge. Be aware that when this happens, both buyer and seller, user and provider, are now continuous learners.

13. Make Sure Your Offer Anticipates Your Customers' Desires

This isn't magic. It's the next step in the feedback loop we talked about in number 12. Once you've equipped your offer to collect volumes of information, enable it to spot the patterns and extrapolate. Your offer builds a history of what a customer wants and how he behaves—always places an order before 9:00 A.M., say, always watches TV programs about travel and cooking, usually buys first-class air tickets, never accepts a hotel room where people have been smoking cigarettes. Your offer should plug such preferences into a customer profile so that you can contact him about a forthcoming no-smokers' gourmet trip to Paris. Similar profiles will help to alert book buyers to the new opus by their favorite author, or sign up digerati for beta tests of forthcoming software upgrades. (You might also be able to sell such information.)

14. Help Your Customers Get Smarter Every Time They Use Your Offer

Some physicians explain the behavior that has brought on health problems, to induce patients to change their behavior. Certain software programs give the users tips on their features, in the hope of educating them to be better users. But the $5,000 commercial stove so popular today doesn't make you a better cook. Why not? When your offer gets smart, you should also be able to learn from it. Your offer must somehow record not only how it is being used but how satisfied your customer was with that use. This enables it to point out changes in circumstances, use, or behavior that would provide the customer with more satisfaction. If your customers are in a business, the offer helps them to do their jobs better; if they are consumers, they get more for their money.

15. Put Filters on Your Information-Intensive Offers

If your offers are connected, and information is flowing through them, soon that will mean too much information flow unless you put filters on them. This is a first cousin to the smartness of the offer.

16. Forget about Annual Model Changes; Download Your Upgrade

When speed is your mind-set, you're operating in real time, online and interactive. It's now possible for your customer to download any upgrades. Your offer, having learned something in the exchange, leads you to bring out newer and better versions. Post them immediately; let your customers click-click and incorporate them into their version. Then, they won't have to trade in to trade up. This also gives the business a continual feedback loop. A Hewlett-Packard printer is a darn smart machine, but it gets even smarter every time a new release of the driver is downloaded from the Web site. When an offer is connected and learns, you can say goodbye to annual model changes, whether they come from Ford in Detroit, from Microsoft in Seattle, or from you. Be warned, however, this will change the economic model of your offer: How do you collect every time your offer improves? But that, too, can BLUR your business to advantage.

17. Extract Information from Every Buy/Sell Exchange

Every exchange you're involved in, whether it's as a seller who's buying or a buyer who's selling, comes with an enormous amount of extra information. Grab it, store it, use it down the road.

18. Buy While Selling

This is one of the cornerstones of blurring your business. The old walls that once separated buyers and sellers are gone. Instead, you should never sell anything without getting something in return. It can be money, but what you're really after is anything and everything you can learn from and about your customers. Yes, it would be great if you could collect this kind of information for free. Fuggedaboudit! Customers have wised up and want to be paid for what they once gave up for free. You're looking at a bargain nonetheless.

19. Sell While Buying

Read number 18 backward. You've got it.

20. Remember: Every Sale Is an Economic, Informational, and Emotional Exchange

This is exactly what it says. Every sale involves the exchange of all three, not just the cash. When you make a sale, be sure you also know what information you want to collect during the exchange and are able to do so. Do the same for emotions. Identify the ones you're selling and the ones you're buying. Start with simple lists; refine them, operationalize them, collect them, and measure the important ones. Develop strategies to amplify them.

21. Put Emotions into Every Offer and Every Exchange

Use the tools of technology to convey sound, motion, color—and, soon, smell—to everyone involved in your economic web. Create a personality with an emotional impact consistent with the customer experience you want, and your whole organization and web of partners will become aligned with it.

22. Put Emotion into Every Other Aspect of Your Business

Extend the emotional experience of your customers to every aspect of your organization. Why should the customer have an emotional reaction different from everyone who makes up your corporate self, down to the factory floor (even if the factory is now making software)? In health care, we expect nurses to have empathy, but have accepted less from our doctors and administrators.

23. You Know the Fortune 500. Get to Know the Value 500

You can name at least 20 of the Fortune 500 companies, right? What about powerful customer or consumer groups, formed or unformed, that will have an impact on your business? Make a list of as many customer groups as you can think of, those that relate to your business today and those that don't. These are the foundations of what we're calling the Value 500 and they will shape the market in the future. Identify which ones will be important to you. What will they care about? Where will their allegiances lie? How do you make them believe in you? Your customers won't be unorganized much longer. Now's the time to figure out how things will change when they come together. If you are among the first to understand this, you can help sponsor this development—help them get off the ground, monitor their concerns, and feed back some answers. By anticipating demands they would make as organized antagonists, you can turn them into fans.

24. Virtual Location, Virtual Location, Virtual Location

We all know the three-word mantra of real estate experts. Well, Connectivity or not, it's great to be literally down the street from your friends, customers, suppliers, and other partners. But Connectivity, not unreasonably, demands that you expand your neighborhood. If it is measured by a 10-minute journey, you can go anywhere on Earth and find the joint hopping 24 hours a day when you get there. Your imperative is to create a community of like-minded players, wherever they are physically. The beauty is that it's easy. Colocate customers and suppliers virtually with an auction market. Colocate collectors—for Beanie Babies, say—with a shared virtual clubhouse. Colocate researchers with an accessible database. Where does your intended community go today? Link to all those places. Show up in front of those people like Coke—six virtual feet away.

25. Forget All You Thought You Knew about Business Economics

Not to sound too pointy-headed, the dismal science has focused on what's called "comparative statics"; that is, what will happen when a small change (in supply or demand, for example) occurs in a competitive marketplace. Cases of noncompetitive markets were called "market failures," meaning they were not the economists' fault, they were real life's fault. The speed of BLUR makes clear that what's important is not the details of the static equilibrium, but the major features of how things are changing. For example, you can't count on competition giving way to consolidation and oligopoly; you need to plan for obsolescence before consolidation is even considered. Or, you can't assume that low levels of personnel turnover are more efficient, when the task is innovation, not production.

26. Forget the Law of Diminishing Returns

An increasing proportion of your business—all the knowledge that is embedded in software—obeys the laws of increasing, not diminishing, returns. That means that as all businesses become more information-intensive they will increasingly migrate toward increasing returns. That also means someone will come to be the dominant provider of it, at rates far cheaper than you can do it for yourself. So, in particular, get ready to outsource your infrastructure and support. Stop writing your own code, making your own product, or training your own people, if someone else is focusing on doing it; and decide which part of the business you will keep as your core.

27. *Don't* Start with Your Customer

The wisdom of the late industrial era was always to start with what the customer needed and backtrack to which products and services those needs called for. That fit when the customer already understood the need and the product, and innovation meant a different shaped bottle for liquid detergent. In BLUR, technical change is happening so fast, your product must educate the customer (beepers for kids on dates?) and the customer must educate you. You can't afford the time delay to put something new in front of the customer. Instead, start with what technology will make possible, codevelop it as fast as you can with the customer, and be flexible and adaptive enough to adjust it according to customer needs as you go. As in software, the first release is your take on things. The customer enters the feedback loop and starts to influence things with release 2.0 and beyond.

28. Don't Grow What You Can Buy

Growth is organic and natural, and you often don't have the luxury of time that this takes. The need for Speed makes it necessary to buy growth instead. Connectivity makes it possible, and Intangibles make it easier. Everything is open to expedience. If you end up as merely the best, fastest orchestrator . . . what's wrong with that? You're the fastest, most intangible, best connected competitor in the Valley.

29. Don't Plan Your Company's Future; Adapt

The faster things move, the less time you have to plan for them. You're much better off iterating and reiterating, adjusting as you go. Bacteria didn't plan people, but they sure are thriving on them.

30. Learn to Partner, Learn to Split

Nothing is forever in the blurred world. This means you must learn to forge partnerships of convenience—with customers, suppliers, and competitors, among others—and then move on when the benefits dry up. You must be on a constant prowl, searching for new partners, sometimes to share in a new offer, sometimes in a new process. The secret to a great relationship: You stick to your core competencies and let your partner (or partners) stick to theirs. Old enemies can make great, temporary bedfellows. IBM and Apple teamed to produce the PowerPC, for example. Even trickier, learn how to recognize when it's over, and call it quits. Timing your goodbye is as critical as that first handshake or make-up kiss. Part on good terms. Goodbyes aren't forever either.

31. Create a Platform; Be the Standard

One of the great enablers of Connectivity in the blurred world is the existence of common standards and platforms. Make it your goal to set a standard—in whatever business you're in. It may mean giving away your first 10,000 units (à la Netscape). Those who succeed in setting the standard in a field (example: Microsoft Windows) stand to profit tremendously. Identify the components of your offer, including the users' knowledge, and find a way to bind them to your standard.

32. Let the Market Price Your Offer

Don't wait for feedback on how your offer is selling to call a meeting on whether to raise or lower prices. Instead, establish a feedback loop that lets the market itself make the changes directly. This is how financial instruments such as stocks, bonds, and other securities are priced all the time. We have all become used to the prices of airplane tickets and gasoline changing daily, but we don't like charges "over list" for a hot car model. Get your customers ready: Before long, you will come to read the price of a head of lettuce in a supermarket on a continuously changing LED screen. Find a way to get price information directly from the market—from the Net or from your existing channels.

33. Let the Market Market Your Offer

The traditional "4Ps" of marketing are price, product, physical distribution, and promotion. Pricing was dealt with in number 32; we've talked a lot about the offer, and physical distribution for Intangibles is not a problem; but what about promotion? The trick is to give away just enough of your offer to get people talking about it. Once the buzz starts, they'll recognize your name when they hear it, will start asking for it, and will be willing to pay your price. Kim Polese, founder of Marimba, a software start-up, went this one better: She let the capital market manage her offer. Her very public venture funding created much attention to get everyone waiting for their then-in-vapor Castanet. Determine who is interested in telling others about your offer, and give them any help they need.

34. Assume Everything Will Be Deregulated

Take this as axiomatic: Technology will change faster than government can change the regulations to address it. Everything about computers and the Internet changes much faster than the policymakers and lawmakers are capable of solonically addressing. Everything about biotechnology is creating ethical dilemmas faster than we can legislate ethical solutions. Even when the lawmakers vote it, the executive branch promulgates it, and the lawyers adjudicate it, there will be more dilemmas than the one they actually addressed. Some regulation is necessary and good. But it is still too slow. Assume your

offer will be judged by the rules of the market, not those of regulation. Create the scenario for how your world will look absent regulation, and then start worrying about how real life will be different.
Don't get in the box of assuming regulation and working backward;
you'll be beaten to the punch by a free agent acting blurred. Don't be
content with the regulators' next incremental change—define what
the market wants and head for that.

35. Measure Your Company by Market Cap Ratios, Not Revenues

We've said all along that it is market capitalization that reflects
future value. So start measuring your performance in terms of market
capitalization. This is not to say you should focus even more patho-
logically on your day-to-day stock price, but that you—even if your
company is privately held—should consider wealth creation as your
company's true test. How does your market capitalization per employee compare with that of other companies making comparable offers
or employing comparable talent? The talent will flow to the companies that offer real rewards, and that means wealth. Today, stock
options are granted only to the senior; tomorrow, they will rival
salary as the primary form of compensation. So, also count the number of millionaires you're creating, in your company and in your web.
Remember, in 1997 alone, Microsoft created 21,000 millionaires, and
that was just among its employees.

36. Churn to Evolve

If you want to be adaptive, innovative, and flexible, you must introduce novel elements into the change. That's what churn is about. If
everybody in your department comes from Ivy League schools, hire
an articulate trade school grad with an electric guitar and a tattoo. If
your inventory is turning fast but always stocking the same stuff,
introduce variety. If 80 percent of your offers are more than a year
old, upgrade 40 percent of them. Get churning.

37. Replace Management Commands with Market Signals

If you can remember only one thing about how to organize, it should be
this: Market signals are the best measure for what to do in your organization; internal power, status, and psychology are the worst signals.
What is dictated by fiat in your company? Capital allocations? Head-
count? Create markets for these resources driven by external signals.

38. Push Power to the Periphery

If your organization has more than a few people, it almost certainly
has some kind of hierarchy—and that means that power concentrates
at the center-top. We know that the shift from mainframes to PCs
has pushed power back down, to the periphery. That's exactly where
you want it to be. But power obeys some weird laws of physics, and
always seems to drift back up and in. Consciously every month, monitor whether that upward drift is happening again. Like weeding a

garden, this is an ongoing task. For operating purposes, you want to put the power where the daily action is and where the decisions need to be taken—at the point of contact between a front-line employee and your customers. This will further help you BLUR boundaries that separate you from your customers. Then set up a process to take management direction from the frontliners.

39. Be Big and Small Simultaneously

What's the right size for your organization? The historical sweep was from small to big. The future will have both at the same time. Small units are nimble, and can act, move, and change swiftly. As long as they are electronically connected, there is no disadvantage of scale. At the same time, these small units must behave and compete globally. So, if you're an individual acting as a free agent, just make sure you're electronically connected to a big economic web. If you're a large corporation, make sure you disaggregate your organization to small work teams of 1 to 35 people. Thus, if you're currently big or small, you're only halfway there. You also need, paradoxically, to simultaneously become the opposite.

40. Tear Down Your Firewalls

Forget the old arm's-length model. Here's the new rule: Tell everybody everything. Yes, you have to protect yourself against electronic vandals, thieves, and various forms of hacker. That's only common sense. But since those who want information about you can easily get it by indirect means, why not plug them in directly? Remember, your advantage is in Speed, and your organization can't act fast if it doesn't know what's going on. By the time your competition sees what you're up to, you'll be long gone.

41. Avoid Maturity

Life cycles have four quarters: gestation, growth, maturity and decline. You've always been told to avoid being a fourth-quarter company. Now, with the rapid pace of change, we're saying to avoid even the third quarter. Reinvent your strategy so that by the time growth turns to maturity, you're already launched onto a business redefinition that has you growing again. There was nothing more mature than the soft drink business. When Coca-Cola invented "share of stomach" to replace the tired market share measure—and said, at best, it had only two of 60-plus ounces in that gustatorial market—it instantly re-created itself as a growth company, liberating incredible new energy and expansion for the company.

42. Blend Webs and Hierarchies

Focus on building your future organization web, not on tearing down your old organization hierarchy. Hierarchy in your organization will only recede to the degree that it is replaced with something better. Perhaps one day webs will replace hierarchy totally, but we doubt it. You're much better off blurring them, rather than pitting them against each other as adversarial models.

43. Use It, Don't Own It. If You Do Own It, Use It Up

In today's virtual world, you should keep your assets in cyberspace and off the balance sheet. The more intangible your business, the easier it will be for you to manage it according to BLUR economics. The more stuck you are in physical stuff—factories, offices, the equipment it takes to run them, payrolls—the more you'll be locked into yesterday's rules.

44. Prize Intellectual Assets Most, Financial Assets Second, Physical Assets Least

Here's the way to go on this one: Add up your assets for each of these three categories. If your intellectual assets aren't the winners, get started on a plan to make them so. If you've got huge value in plant and equipment that's worth much more than your people, then your strategic chore is to answer how you'll move rapidly in the opposite direction. Determine the crossover point. If it's in the invisible future, then it's likely someone will come along and eclipse you. What? You don't know how to add up your intellectual assets? Move on to number 45.

45. Manage, Measure, and Grow Your Intangible Capital

Buy-side security analysts attribute 35 percent of market value of the stocks they follow to intangible variables not covered anywhere on financial statements.[1] Examples include the reputation of an R&D staff in the case of pharmaceuticals, or management experience in the computer industry, or brand and distribution in oil and gas. Determine how much of your market capitalization is determined by such factors and how you score on these dimensions compared to others. Where you're good, tell the financial markets! Where you're not, find out why and do something about it. The accounting profession will learn to measure these Intangibles eventually, but meanwhile, start assessing this increasingly important dimension of value.

46. Value a Company by Its Growth Rates, Not by Its Assets

The financial markets determine price/earnings ratios in large part based on the expected growth rate of earnings. Yet the accounting rules focus on the market value of assets. To see the value your business is creating, develop a dashboard that measures your acceleration. Determine: How much faster are earnings growing than last year? What about profit per employee? Perhaps most important, are you paying attention to the growth rate of your employees' wealth?

47. Own the Links, Not the Nodes

Remember you're living in a world where everything is connected with everything. So master the ability to Connect. Today's node may churn and change but you'll still want to be connected to whatever has replaced it. We already tore down the firewalls back in number 40. Now train your people to network, give them the tools they need, and allocate time to relationship planning and building. Contents change; conduits go on forever.

48. Value What's Moving, Not What's Standing Still

Anything standing still is a liability that's using your money unproductively, whether it is a piece of inventory, a nonproductive employee, or a pot of gold under your bed. Motion makes them productive, so you've got to know how to measure, manage, and value Speed more than quantity. How fast does your inventory of ideas turn over? How do you know?

49. Get Ready for Triple-Entry Bookkeeping

We keep talking about being future-oriented, and that value is measured by future prospects. These in turn depend on growth. Use your measurement system to track growth and react to changes in it. In other words, manage acceleration, not Speed itself. Small companies with high growth rates create incomprehensible earnings-per-share multiples, while well-established, large, slow-growing companies languish. This applies to all kinds of measurements, not just sales and profits. Growth of scarce capabilities, connections to your web, products introduced all have to increase as the BLUR accelerates.

50. Pay Attention. Attention Is the Next Scarce Resource

We live in a world of information glut. Start paying attention to how to get people's attention and keep it. We're not talking about a cute gimmick. You must get serious about it. Be warned: As you get serious, you'll start having meetings and writing memos, and ultimately, getting attention will be a topic for bureaucracy. The cure will be worse than the disease, and the enemy will be us. Getting and keeping attention is serious, but it has to be lighthearted and engaging. This has to be true of the office and the store.

These 50 should be enough to keep you busy with your business for quite a while, but be sure to save some time for . . .

10 Ways to BLUR Yourself

1. BLUR the Divide between Work Life and Life Life

Do you work to live? Or live to work? In BLUR, these aspects of life aren't separate. Everyone knows the line is already indistinct, as voice mail invades late nights, flights shorten the weekend at both ends, and pagers disturb you in the theater Saturday night. There goes the rest of another weekend. Chris has even received a phone call from a gondola in the Grand Canal; and at a party in Budapest, Stan's Indonesian friend took out a near–cigarette lighter–sized cell phone on which he could be reached anywhere in the world (except the United States and Japan!).

So, are the BLUR brothers telling you to disconnect? Far from it; we want you to go with the BLUR. Run your home by the BLUR, too, and you won't see work and life in a disharmonious relationship, but as a better integrated, better organized world. Use Peapod to shop for groceries, Amazon.com to hunt for books; e-mail your friends; check the Web before you check the Yellow Pages. It's working for business, it can work for you too. And, make your work life more "livable." Let the home invade the office, too: Bring the kids, flex the hours, wear what makes sense for your business. It will help generate the emotion and relationships your business needs.

2. Have Your Cake and Eat It Too

What? Has every schoolchild's dream come true? In the world of BLUR, yes. Remember, knowledge, a main currency of the BLUR economy, has a unique property: You've got it, you sell it—you've still got it! So, don't hoard your knowledge. Spread it, get credit for having known it early, become known as the source of interesting ideas, whether they're original or secondhand. If you don't, your friends will hear it from someone else. Velocity of knowledge is crucial to your success. The more you give away, the more you'll get back.

3. Seek Novelty Forever

Agency.com is an online ad agency, more or less. Here is the mission statement from founder Kyle Shannon and his management team: "Agency.com is a group passionately committed to creating something out of nothing every single day forever." What about you? Remember, in BLUR, your skills are obsolete before you wear out, life expectancies are lengthening, but social security isn't. If you're not creating something new all the time . . . (we leave the ending of this to your imagination).

4. Moonlight from Strength

PreBLUR, you moonlighted only when you needed extra money. Now that you're managing your own stock price, you can't afford to dedicate yourself solely to the needs of your organization morning, noon, and night. Car cell phones and e-mail that reaches you at home on the weekends are both helping to stress you and everyone else to the max. If you stay caught up in this, you'll always be beating back the encroachments like a brush fire that won't be extinguished. You'll be much better off, psychologically as well as financially, when you apply BLUR rather than get blurred. Diversify. Moonlight because it rounds you out, earns you new experiences, tells you first-hand about what's out there, spreads the word of your value. Get a sabbatical, work for a customer, take a course, build other relationships. You'll be doing your employer a favor and increasing organizational churn and diversity, so don't hesitate to get paid for these activities.

5. Sell Your Value on the Web

A fine place to spread the word of your worth is on the Web, a great leveler of the meek and the mighty in that it gives an equal shot at a global market to both you and the giant corporations. In just a few hours and with only a fistful of dollars, anyone can put their offer on the Web, and cast for buyers. Although it's still not easy to transact those offers for dollars, now is the time to climb aboard the learning curve. You want to gain experience in the virtual marketplace, to prepare yourself for the economic-information-emotion exchange. Don't try to land a one-month full-bore job on the Net the first time out. Rather, dip in your toe, experience exchanging yourself in cyberspace. Establish your market, however small it seems at first. You are your offer.

6. Let the Market, Not the Company, Determine Your Worth

Let's assume you've either landed that great promotion you went after or you've moved to another company for a better position. Either way, somebody has recognized your true abilities. The point to remember is that the market doesn't lie. It's a far better arbiter of what you're really worth than all but the most perceptive of bosses. Generally, companies have much more restrictive ceilings on your future than does that much larger market out there. Companies have budgets, guidelines, Hay Points, policies, rules and regulations. The market doesn't. It just looks for value and is fully prepared to pay for it. So even if you want to stay where you are and plan to put in some time there, be sure you use the market as the yardstick by which to set your compensation package.

7. Become a Free Agent While Still on a Payroll

This one is really a mind-set change. You have to stop thinking of yourself as a wage slave—which is how many companies think of their employees—and wake up to your new freedom. For a newly emancipated slave, this way of thinking carries new responsibilities: You, not your boss or employer, are responsible for your future. In other words, even though you're on somebody's payroll, think of yourself as self-employed and currently under contract to one team. You may stay with that team forever, or you may move on every season. The point is not how peripatetic you are, but what's going on inside your head. Paternalism ended decades ago, employee-ism is fading, free agentry is how we'll work in the BLUR of tomorrow.

8. Brand Yourself; There's Equity There

As a free agent, with a value to market, brand it! Writers talk about "finding their voice," which means they've hit their stride and their words are distinct from everyone else's. You need to have a distinctive brand—a voice—too. This doesn't mean to become distinctive just within your department, either. Think big: Be recognizable inside and outside your immediate area of work. When your brand sells in the *market*, out there, you have the potential for true brand equity. Martha Stewart could still be a housewares buyer in a department store; instead she's a cult, broad enough to inspire a parody magazine (*Is Martha Stewart Living?*).

9. Securitize Yourself

You already know that intellectual capital—your intelligence—is your most valuable resource. So capitalize on it, literally. This means take yourself public, just as companies do. Famous sports figures and entertainers are already starting to do this successfully, but you don't have to wait for the trend to trickle down. One way to begin is to sell a percent of your future earning stream to a bank in the form of a lien, or to your employer as a piece of a private company; in this case, You, Inc. If your employer is not yet blurred enough to capture the concept and run with it, offer such shares to friends, colleagues, family, and other people who know and admire you. You can even offer shares to competitors. (That will get your boss's attention!) A few years down the road, when securitization of individuals becomes common practice, as it will, professional investors will move in.

They'll be asking you questions about your plans and projected earnings, and then buying and selling your shares publicly. Always remember to keep at least 51 percent, which keeps you in control of your own destiny. Also be aware, there'll be times when you or your company will want to buy back your stock.

10. Manage Your New Dual Career

Whether you've securitized yourself or not, you should be aware by now that you've really got two careers, not just one. Traditionally, ambitious people have made very careful career plans that have encompassed strategic promotions and potential moves from one company to another. Now it's time to wake up and pay just as careful attention to your ascent out there in the financial marketplace as you do to your ascent inside a company. Before, when you entered the free market, you probably did so only as a stop gap between jobs. We're telling you to always do both simultaneously: Treat the market and the company at which you're on a payroll as two places where you're doing your thing. Hasn't every boss you ever had told you to be entrepreneurial? You have to take that a step further and do it literally. Play the game inside and outside your company. In other words, take both balls and run with them. This is your true chance to act like the free agent you are.

The 60 suggestions that made up this chapter are a beginning—our ideas on ways you can embrace the BLUR. We offer them, as we do all of the perspectives in the book, as the starting point for a conversation. We have tried to be comprehensive, covering the entire span of business, rather than be detailed or exhaustive. We've also tried to be provocative and future-oriented, rather than focusing on how-to and next steps. We hope we have provided you with a lens through which to view the effects of Connectivity, Speed, and Intangibles in your world, and anticipate their implications. Now we'd like to know what you are seeing through that lens, so that we can bring you into a community of BLUR. Here's what we're doing.

Through our continuing research and work with clients, we will be seeing more evidence of BLUR. We intend to document it on the BLUR Web site, www.blursight.com. We invite you to check in frequently to keep up with the BLUR; when you do, you will find the ingredients of your own BLUR scrapbook—a growing collection of examples and links to other sites, and worksheets for collecting your own observations, corporate or personal. You will also find people to talk with who have various kinds of expertise. We ask in return that you contribute your observations, for the benefit of the community, as well as for a reward you'll learn about on the Web site. If you do, we're confident the community will grow, to our mutual benefit, and that it will contribute to the customized vision of BLUR you develop for yourself.

We've finished *BLUR*, the book. Now—better equipped—it's time for all of us to get on with the BLUR in our work and in our lives.

We said at the start that *BLUR* isn't just a book. Like any good offer, it needs to be interactive, upgradeable, and customizable. More than anything, we want it to get smarter with use. For that to happen, we need your help. And we're willing to reward you.

Please give us your BLUR-inspired ideas. We'll choose one each month in 1998 to receive a prize: a check for $100. And those that strike us as most useful to the broadest audience will be published to our Web site so the whole BLUR community can be part of the exchange. (You'll also find opportunities at that site to participate in online discussions and to access more detail on cases and concepts cited in the book.)

Of course, the form is available electronically, at www.blursight.com, or you can tear out the one here and mail or fax it to us. Also, feel free to offer your input by e-mail.

BLUR
The Ernst & Young Center for Business Innovation
One Cambridge Center
Cambridge, MA 02142

Fax: 617-761-4144
E-mail: blur@ey.com

TELL US SOMETHING
WE DON'T KNOW!

What's the best example of BLUR**ring you've seen?** *(We'll love it even more if it's from actual experience—from your business or your life.)*

_____

_____

_____

_____

_____

_____

_____

_____

Reflecting on that example, what's the general lesson to add to "50 Ways to BLUR** Your Business . . . And 10 Ways to B**LUR** Yourself"?**

_____

_____

_____

_____

_____

_____

_____

Complete the following lines if you'd like to be eligible for our monthly prize. *(Or attach your business card.)*

Name _____

Company _____

Role or title _____

Phone _____

E-mail _____

Street address _____

City, State, Zip _____

ACKNOWLEDGMENTS

BLUR was created true to the principles of Connectivity Speed, and Intangibles. The ideas, of course, are inherently intangible, though our efforts were focused on combining conceptual breakthroughs with practical applications. Many people worked on the project in a hyper-connected way; their roles always overlapped and boundaries within the team scarcely existed. Also, unlike the typical cadence of the publishing world for business books, we operated at BLUR speed and went from brainstorm to bookstore in less than a year.

Many of the leading practitioners in Ernst & Young's consulting practice contributed their experience, examples, and reactions to this book. Terry Ozan and Chris Christensen continually extended our thinking with their own, and provided strong sponsorship for the effort. Jim Andersen, Terry Boyle, Phil Fernley, Charlie Gottdiener, John Jordan, Sean Kenny, Peter Novins, John Parkinson, Ralph Poole, Mike Powers, Jim Searing, and Neil Smiley were especially generous with their time. Their view from the front lines provided an important reality check for the research, writing, and editing team.

The editing team, led by Julia Kirby, included Colin Leinster, Ed Wakin, and Andrea Mackiewicz, each of whom added important creative and editorial assistance. Jim Park, David Greene, Steve Johnson and his CBK team, and Sanjeev Acharya assisted Nikolas Kron in the discovery and researching of examples of BLUR, without which our ideas would be only assertions.

The entire staff of the Center for Business Innovation provided ideas and availability to test trial balloons, and stayed alert to developments that had relevance to our work. In particular, Rudy Ruggles and Teresa Parker joined Nikolas Kron in the development of the characteristics of offers. Barbara Elstein did important work underlying the discussion of the exchange. Karen Frasca and Jennifer Cline provided the endless emendation capability required to capture the contributions of such a large and diverse group.

Meanwhile, Avery Hunt, from Ernst & Young's national office, coordinated closely with David Goehring's wonderfully cooperative team at Addison-Wesley, led by Nick Philipson, Elizabeth Carduff, and Pat Jalbert, who, surprisingly for us, actually applauded our insistence on speed. Manuel King, our illustrator, was equally responsive. Finally, our agent Rafael Sagalyn once again skillfully guided us through all the intricacies of the exchange.

NOTES

Chapter 1

1. This was the foundation of Stan's influential book, *Future Perfect* (Addison-Wesley, 1987, 1996).

2. Frances Cairncross, *The Death of Distance* (Boston: Harvard Business School Press, 1997).

3. Ibid. Other sources believe voice will fall to 10 percent in this period. It hardly matters—what is clear is that data traffic is growing at a rate far exceeding that of traditional telephone use.

Chapter 2

1. "Mass customization" was first coined in Stan Davis, *Future Perfect* (Reading, Mass.: Addison-Wesley, 1987, 1996).

2. Today, MacConnection goes one better than Dell: Order by midnight, receive the next day before noon.

3. Ernst & Young research study, 1997. See Stan Davis and Chris Meyer, "They've Already Got Us Surrounded . . . What Happens When They Team Up?," *Forbes ASAP*, June 1, 1998.

4.   Kevin Kelly, "New Rules for the New Economy: Twelve Dependable Principles for Thriving in a Turbulent World," *Wired*, September 1997, pp. 140–197.

5.   Nicholas Negroponte, *Being Digital* (New York: Knopf, 1995).

6.   Todd Lappin, "The Ultimate Man-Machine Interface," *Wired*, October 1995, pp. 124–188.

Chapter 3

1.   The experience of RW Frookies, Inc., referenced by INC. Online, "Cookie Monsters," by Paul B. Brown, February 1989.

2.   *See* Feargal Quinn, *Crowning the Customer* (Dublin: O'Brien Press, 1990).

3.   Cookies were named for a UNIX "fortune cookie" program that output a different message or "cookie" each time it was used.

4.   Ronald B. Lieber, "Selling the Sizzle," *Fortune*, June 23, 1997, p. 80.

5.   Another Internet example: Dozens of day care centers have installed cameras that take photos of their classrooms every minute or so. Parents with the right password can take quick looks at their kids—an emotional value, indeed, as well as a welcome break during the parents' working day.

Chapter 4

1.   Stephen Haeckel, "Adaptive Enterprise Design: The Sense-and-Respond Model," *Planning Review*, May–June 1995.

2.   Cited in Stuart Kauffman, "The Evolution of Economic Webs," in P.W. Anderson, K.J. Arrow, and D. Pines (Eds.), *The Economy as an Evolving Complex System* (Reading, Mass.: Addison-Wesley, 1988), pp. 125–146.

3. David Lane, Franco Malerba, Robert Maxfield, and Luigi Orsenigo have exposed this idea in their work at the Santa Fe Institute. See for example, "Choice and Action" (Santa Fe Institute working paper, 1995).

4. The term "co-opetition" was popularized by Adam M. Brandenburger and Barry J. Nalebuff in their book *Co-opetition: A Revolution Mind-set That Combines Competition and Cooperation: The Game Theory Strategy That's Changing the Game of Business* (New York: Doubleday/Currency, 1996). They credit Ray Noorda, the founder of Novell, with coining it.

5. Kevin Kelly, *Out of Control: The New Biology of Machines, Social Systems, and the Economic World.* (Reading, Mass.: Addison-Wesley, 1995).

6. John Donne, *Meditation–XVII, Divine Poems* (1607), or *Holy Sonnets* (1618).

7. Kevin Kelly, "New Rules for the New Economy: Twelve Dependable Principles for Thriving in a Turbulent World," *Wired*, September 1997, pp. 140–197.

8. From the term that economists give to how markets actually work versus hypothesized perfect competition.

9. *The New York Times* reported on one example of this in its August 31, 1997 issue: Robyn Meredith, "Buick Bringing Its Rivals to the Showroom Floor," Section 1, p. 35, col. 2.

10. Tobin's Q is the ratio of the market value of a firm to the replacement value of its physical assets.

11. "The Fiber Baron," *Manager's Journal*, WSJ-Interactive Edition, October 6, 1997.

12. "Dreary Days in the Dismal Science," *Forbes*, January 21, 1991, p. 68.

Chapter 5

1. In physics, methods such as a finite element analysis and simulated annealing are used to estimate the behavior of large connected systems. These approaches treat each element (say, a small piece of a strut supporting an airplane wing) as an independent element that reacts to incoming stimuli in ways that feed back to its neighbors—a passive but connected agent. New methods coming out of the study of Complex Adaptive Systems are using more active agents, with great "decision-making power." Soon, these agent-based approaches will be suitable for modeling connected networks of active economic agents including firms and consumers.

2. Trees and twigs are alike because they are designed according to the same principles of fluid flow and nutrient requirements at all levels. And here's another interesting similarity: Cardiovascular systems, respiratory systems, plant vascular systems, river systems, and insect-breathing tubes are all examples of fractal branching networks, each tuned to the energy requirements of the hosts they serve. "When it comes to energy transport systems, everything is a tree," says Geoffrey West, a physicist at Los Alamos National Laboratory. Interestingly, Brandenburger has described the growing connecting infrastructure as "the circulatory systems of an artificial life form."

3. Michael Rothschild, in his excellent 1995 book *Bionomics: The Inevitability of Capitalism* (New York: Henry Holt & Company), developed the metaphor of economy as ecosystem in detail.

4. Kevin Kelly, "New Rules for the New Economy: Twelve Dependable Principles for Thriving in a Turbulent World," *Wired*, September 1997, pp. 140–197.

5. Stephen Haeckel explores what he calls the sense-and-respond model for organizations in "Adaptive Enterprise Design: The Sense-and-Respond Model," *Planning Review*, May–June 1995.

6. This experience was presented in an Ernst & Young conference entitled "Embracing Complexity," and is summarized in a proceedings document by that name. Ernst & Young Center for Business Innovation, 1997.

7. Robert Eccles and Dwight Crane, *Doing Deals: Investment Banks at Work* (Boston: Harvard Business School Press, 1988).

8. This term comes out of the work of Per Bak; see Per Bak and Chen Tran, "Self-Organized Criticality," *Scientific American*, January 1991, pp. 46–54.

9. Stuart Kauffman, *At Home in the Universe* (New York: Oxford University Press, 1995).

10. Stephen Jay Gould, *Full House: The Spread of Excellence from Plato to Darwin* (New York: Random House, 1996).

11. Specific market and offer data have been disguised.

12. One of these prefigured BLUR: He recognized that footspeed was the only baseball skill valuable on both offense and defense.

13. Kevin Kelly, *Out of Control: The New Biology of Machines, Social Systems, and the Economic World* (Reading, Mass.: Addison-Wesley, 1995).

Chapter 6

1. The $55 million issue of 10-year notes was bought in its entirety by Prudential Insurance Company of America. The notes, which are rated single-A-3 by Moody's Investors Service, have an interest rate of 7.9 percent.

2. Adam Bryant, "Managing Dilbert Inc.," *The New York Times* Sunday business section, September 7, 1997, p. 1.

3. See "Music: Wall Street Bets on Entertainment Idol's Earning Power" by Bruce Orwall, *The Wall Street Journal*, Friday, September 26, 1997, p. B1.

4. "The Market is Bullish on Michael, But Things Look Sticky for Rice," *The Wall Street Journal*, November 19, 1997, p. B1.

5. Barnett Banks' innovative perks were profiled along with other firms' in the *USA Today* article "Perking Up Employees," by Stephanie Armour, October 8, 1997, p. 1B.

Chapter 7

1. The Egyptians used slavery and the Romans conscription to accumulate capital. We're just Eurocentric here for illustrative purposes.

2. The fact that three books by the name *Intellectual Capital* all came out within a three-month period in 1997 is one indication of the pervasiveness of this idea. The three are by Tom Stewart, a member of *Fortune*'s senior board of editors, Leif Edvinsson, and Michael Malone (Edvinsson is director of Intellectual Capital at Skandia, a Swedish-based insurance firm, and Malone is a noted journalist).

3. Again, we are not attempting a full economic history. Many institutional changes facilitated the rise of debt-to-equity ratios.

4. Yugi Ijiri, *Momentum Accounting and Triple-Entry Booking*, American Accounting Association, 1989.

5. Reva Basch, "Life Online," in *Beyond Cyber Punk*, Hypercard Stack, Gareth Branwyn, 1991.

Chapter 8

1. "Measures that Matter," Ernst & Young Center for Business Innovation, 1997.

INDEX